Joseph Abruscato Joan Wade Fossaceca Jack Hassard Donald Peck

HOLT SCIENCE

Holt, Rinehart and Winston, Publishers
New York · Toronto · Mexico City · London · Sydney · Tokyo

THE AUTHORS

Joseph Abruscato
Associate Dean
College of Education and Social Services
University of Vermont
Burlington, Vermont

Joan Wade Fossaceca
Teacher
Pointview Elementary School
Westerville City Schools
Westerville, Ohio

Jack Hassard
Professor
College of Education
Georgia State University
Atlanta, Georgia

Donald Peck
Supervisor of Science
Woodbridge Township School District
Woodbridge, New Jersey

Cover photos, front: David Muench/The Image Bank; back: Walter Iooss, Jr./
The Image Bank.
The photo on the front cover shows Cape Kiwanda on the Oregon coast. Another section of the Pacific coast, near Monterey, California, is shown on the back cover.

Photo and art credits on page 287

Printed in the United States of America
ISBN: 0-03-003078-1

6789 071 7654

ACKNOWLEDGMENTS

Teacher Consultants

Armand Alvarez
District Science Curriculum Specialist
San Antonio Independent School District
San Antonio, Texas

Sister de Montfort Babb, I.H.M.
Earth Science Teacher
Maria Regina High School
Uniondale, New York
Instructor
Hofstra University
Hempstead, New York

Ernest Bibby
Science Consultant
Granville County Board of Education
Oxford, North Carolina

Linda C. Cardwell
Teacher
Dickinson Elementary School
Grand Prairie, Texas

Betty Eagle
Teacher
Englewood Cliffs Upper School
Englewood Cliffs, New Jersey

James A. Harris
Principal
Rothschild Elementary School
Rothschild, Wisconsin

Rachel P. Keziah
Instructional Supervisor
New Hanover County Schools
Wilmington, North Carolina

J. Peter O'Neil
Science Teacher
Waunakee Junior High School
Waunakee, Wisconsin

Raymond E. Sanders, Jr.
Assistant Science Supervisor
Calcasieu Parish Schools
Lake Charles, Louisiana

Content Consultants

John B. Jenkins
Professor of Biology
Swarthmore College
Swarthmore, Pennsylvania

Mark M. Payne, O.S.B.
Physics Teacher
St. Benedict's Preparatory School
Newark, New Jersey

Robert W. Ridky, Ph.D.
Professor of Geology
University of Maryland
College Park, Maryland

Safety Consultant

Franklin D. Kizer
Executive Secretary
Council of State Science Supervisors, Inc.
Lancaster, Virginia

Readability Consultant

Jane Kita Cooke
Assistant Professor of Education
College of New Rochelle
New Rochelle, New York

Curriculum Consultant

Lowell J. Bethel
Associate Professor, Science Education
Director, Office of Student Field Experiences
The University of Texas at Austin
Austin, Texas

Special Education Consultant

Joan Baltman
Special Education Program Coordinator
P.S. 188 Elementary School
Bronx, New York

CONTENTS

SKILLS OF SCIENCE

Have you ever seen a nest like this? How can you find out what kind of bird made it? One way is to watch the nest closely. Perhaps a bird will land on it. Another way is to find out what the nest is made of. Where did the bird find the things to build this nest? If you can find the answers to these questions, you will know more about the nest and the bird that made it.

Do you want to know more about things that you see around you? Do you try to find out more about these things? If so, you are acting like a scientist. A scientist is always asking questions and looking for answers.

In order to find out what made the nest in the picture on the left, you could **observe** it. When you *observe* something, you study it carefully. If you observe the nest, you will learn that it was made by a red-winged blackbird.

Observe: To study carefully.

Nikolas Tinbergen is a scientist who studies animals. He has learned a lot about animals by observing them. One of the animals that he has studied is the digger wasp. This wasp lives in a tunnel in the sand.

Tinbergen observed a digger wasp for many days. He learned that it digs its tunnel with its front legs. It lays eggs in the tunnel. The digger wasp spends the day hunting for bees, which it feeds to its young.

Tinbergen noticed that it seemed to be easy for the wasp to find its tunnel. He wanted to discover how this happened. Then, he had an idea. Perhaps the wasp used objects on the sand as landmarks to find the tunnel. Tinbergen had observed that the wasp always made a circle over its tunnel before it flew away. Perhaps it checked for landmarks.

Tinbergen decided to test his idea. One day after a wasp left its tunnel, Tinbergen brushed away the pebbles and other objects that were near the opening. Then, he watched. When the wasp flew back to its tunnel, it could not find the opening. The wasp looked and looked. It took 25 minutes before it found its tunnel. Tinbergen had observed that the wasp took only one second to find its tunnel when the landmarks were in place. How much longer did it take the wasp to find its tunnel when there were no landmarks?

How did Tinbergen find the answer to his question? He found the answer by observing the wasp and by testing an idea.

hand lens

field glasses

clock

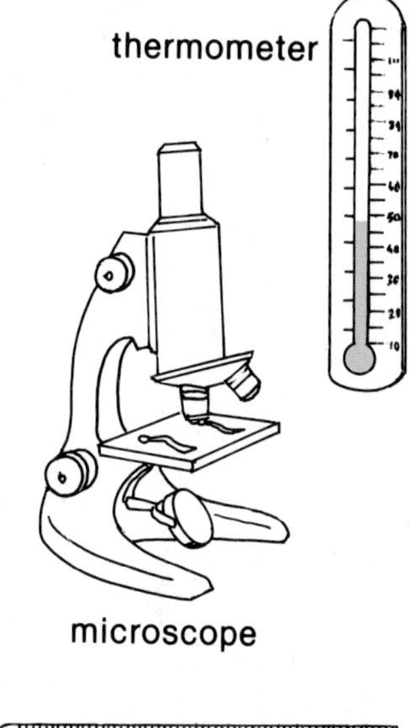

thermometer

microscope

Scientists sometimes use special tools in their work. Field glasses, hand lenses, and microscopes are tools that make things look bigger. Scientists also use tools for measuring. Do you know what can be measured with a thermometer? What can be measured with a metric ruler? What can be measured with a clock?

ruler

SKILL BUILDING ACTIVITY

Does a familiar task take more or less time than an unfamiliar task?

A. Gather these materials: paper, pencil, and watch (or clock) with a second hand.

B. Write your name with your right hand. Have someone time you. Record the time in seconds.

C. Using your left hand, repeat step B.

1. Did it take more or less time to write your name in an unfamiliar way? Explain your answer.

Compare: To tell how something is like or different from something else.

Pretend you saw an animal that you had never seen before. How could you tell another person about this animal? One way would be to **compare** it to an animal that you had seen before. When you *compare* things, you tell how they are alike and how they are different. Comparing one thing with another is an important part of observing.

SKILL BUILDING ACTIVITY

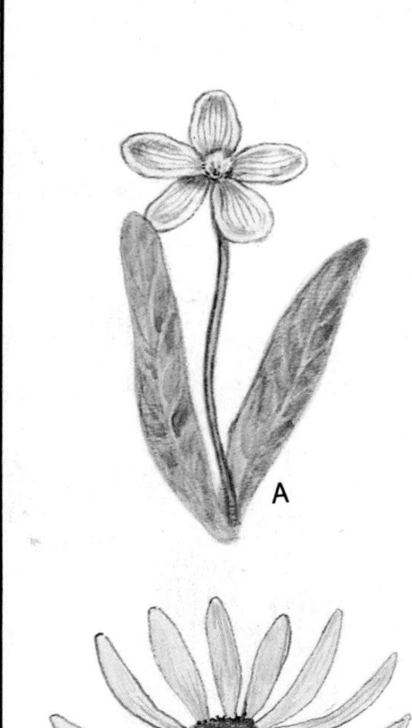

How are these flowers alike? How are they different?

A. Gather these materials: paper, pencil, and metric ruler.

B. Make a chart like this:

Observations	Flower A	Flower B
Color		
Number of petals		
Shape		
Width		

C. Look closely at these pictures. Then, complete the chart.

1. Write a paragraph that explains how these flowers are alike and how they are different.

It sometimes takes hours, days, or even years for scientists to find the answers to questions. Do you know how they remember what they observe and learn each day? They **record,** or write down, their observations in a book called a diary. *Recording* the results of observations and tests is an important part of a scientist's work.

Record: To write down.

Some of the steps a scientist takes to find answers to questions are recorded here. Can you name these steps?

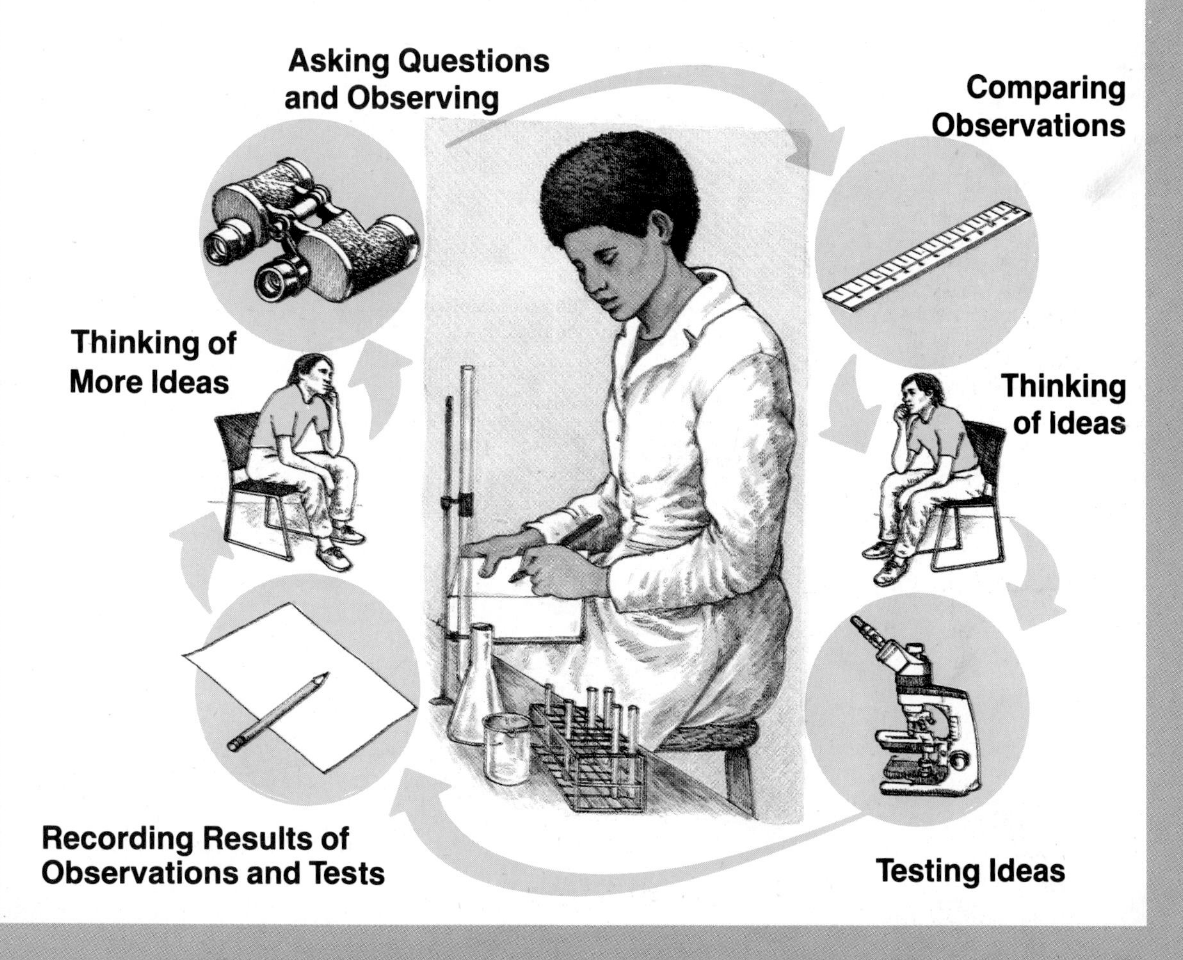

THE LIVING WORLD

UNIT 1

CHAPTER 1

WHAT IS A LIVING THING ?

1-1. Living and Non-living Things

Something in this yard runs, jumps, and barks. What is it? Is it a living thing? Something else moves backward and forward. It squeaks and clanks. What is it? Is it a living thing?

When you finish this section, you should be able to:

- ☐ **A.** Describe what living things do.
- ☐ **B.** Explain how living things are different from non-living things.
- ☐ **C.** Explain how plants and animals are alike and how they are not alike.

Many things around you are living. Other things are non-living. Look at the picture on page 8. A strong, healthy dog is in this backyard. It is easy to tell that a dog is a living thing. A dog can move by itself. The swing in the picture can also move. But it cannot move by itself. A swing is a non-living thing.

Do you know why animals must eat? Young animals need food to grow and to stay alive. Grown animals need food, too.

Does a swing need food? Does a swing grow? The answers to these questions are no. Now, ask these questions about other things in the picture on page 8. If the answers are yes, these are living things.

Do the plants in this picture need food? Yes, they are also living things. Green plants make their own food. Some of that food is stored in the leaves and stems of plants. Food is also stored in the roots and seeds. Some animals eat these parts of plants.

Look at the mother bear and her cubs. Do the cubs look like their mother? All living things can **reproduce** (ree-proh-**doos**). That means they can make more living things like themselves. All animals and plants can *reproduce*.

Reproduce: To make another living thing like itself.

The pictures on these two pages show living and non-living things. Can you tell what is living? What is non-living?

A car uses gas. In some ways, gas is like food. Cars need gas to run. Is a car alive? No, a car cannot grow. A car cannot make another car.

Some icicles appear to grow. One by one, drops of water move to the end of an icicle and freeze. Slowly, the icicle becomes bigger. Does this mean the icicle is a living thing? Can an icicle move by itself? Does it need food? Does it reproduce?

Cars and icicles are non-living things. They cannot do everything that plants and animals do. Now, let's see if plants can do all the things animals can do.

ACTIVITY

Can the leaves of a plant move?

A. Gather these materials: small potted plant and jar of water.

B. Set the plant a few feet from a window.

C. Notice the direction the leaves are facing.

D. Observe the plant each day for a week.

1. Did the leaves move? Explain.
2. How does this activity show that a plant is a living thing?

E. Be sure to remember to water the plant as needed during this activity. An adult can help you decide when the plant needs water.

You know that plants need food to grow. Plants also reproduce. Have you ever planted seeds for vegetables or flowers?

Now, think of ways plants are not like animals. Can plants make sounds? Do they have legs? Do they see or hear? Can they move by themselves?

Study the pictures of the morning glory on page 12. The picture on the left shows the plant in the morning. The picture on the right shows the plant at night. What does the plant do in the morning? What does it do at night?

Can you think of other ways a plant can move?

Section Review

Main Ideas: The world is filled with living and non-living things. This chart shows how plants and animals are alike.

Type of Living Thing	Needs Food	Produces Own Food	Grows and Changes	Repro-duces	Moves by Itself
Green Plant	Yes	Yes	Yes	Yes	Yes
Animal	Yes	No	Yes	Yes	Yes

Questions: Answer in complete sentences.

1. Why do animals need food?
2. How do you know a car is not alive?
3. Where do plants store food?
4. How are animals different from plants?
5. What does reproduce mean?

1-2.

What Are Living Things Made Of?

You know that an onion is round. It grows in the ground. An onion has a strong taste and a strong smell. Some onions are white. Others are red. Do you know what an onion is made of?

When you finish this section, you should be able to:

- **A.** Describe a *cell*.
- **B.** Explain how plants and animals are alike and not alike.

To find out what an onion is made of, you could peel a thin layer of its skin. Then you could look at this layer under a **microscope** (my-kruh-skope). A *microscope* is a tool that makes small things look larger.

Microscope: A tool for making small things look larger.

This picture shows how a piece of onion skin looks under a microscope. You can see that an onion is made of little boxes, or blocks. Each of these tiny boxes is called a **cell**.

All living things are made of *cells*. An onion has thousands and thousands of cells. People have even more cells.

The drawing shows what a plant cell looks like. Notice that the cell has several parts.

A skin-like covering called a **cell membrane** (**mem**-brayn) surrounds the cell. Next to the *cell membrane* is a thicker covering. It is called a **cell wall**. Both coverings hold a cell together and give it shape.

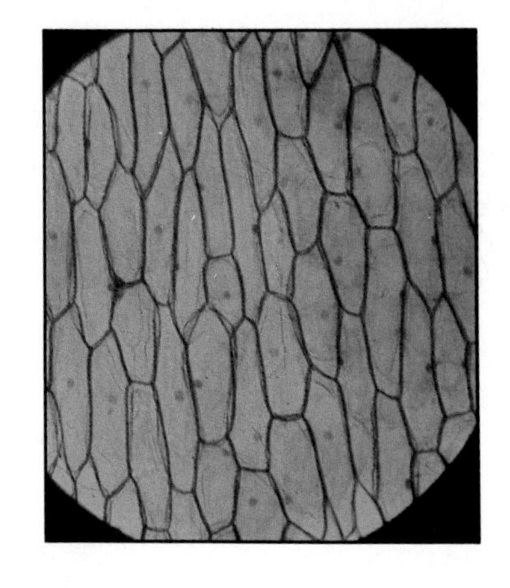

Cell: A small unit of life.

Cell membrane: The thin, skin-like covering that surrounds a cell.

Cell wall: The hard outside covering of a plant cell.

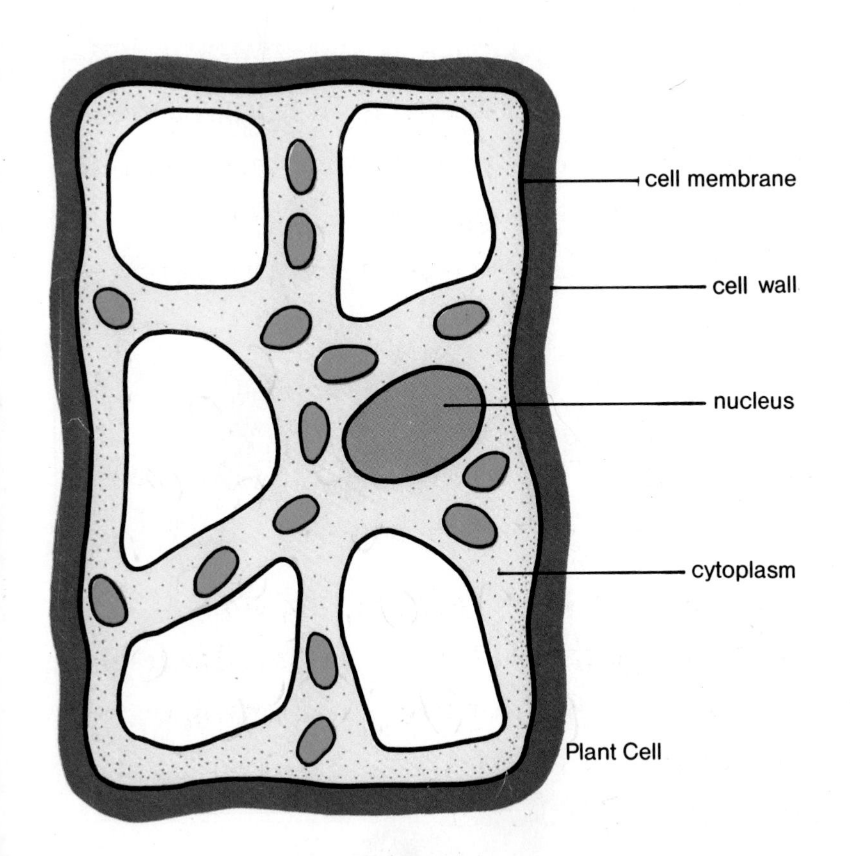

Plant Cell

ACTIVITY

Can you make a model of a cell?

A. Gather these materials: white and blue modeling clay, jelly beans, and knife.

B. Using the white modeling clay, make a round shape about the size of a baseball.

C. Then, using the blue clay, put a thin covering over the ball you made in step B.

1. What does the white clay represent?
2. What does the blue clay represent?

D. Cut the ball and a jelly bean in half. Have an adult help you. Then press half of a jelly bean into the center of each part.

3. What does each half of the jelly bean represent?

A cell is filled with a jelly-like liquid. This liquid is called **cytoplasm** (**sy**-toh-plaz-um). *Cytoplasm* moves inside a cell. Little pieces of living matter float inside the cytoplasm. The biggest piece is called the **nucleus** (**noo**-klee-us). The *nucleus* controls what the cell does. If a cell were a boat, the nucleus would be its captain.

Cytoplasm: The living liquid inside a cell.

Nucleus: The control center of a living cell; found in the cytoplasm.

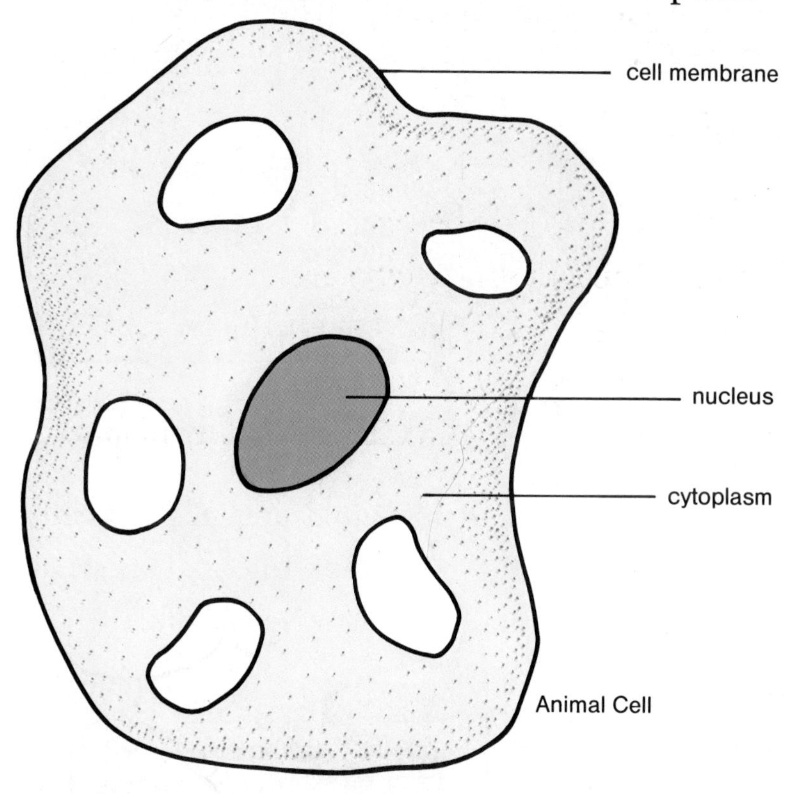

A cell can grow. A cell can use food. A cell can reproduce. The nucleus controls every one of these activities.

Most cells have a nucleus, cytoplasm, and a cell membrane. But that does not mean that all cells are exactly alike.

The above is a picture of an animal cell. How is it like the cell of an onion? How is it different?

Plant cells are stiffer than animal cells. What does a plant cell have that an animal cell does not have? A plant cell has a cell wall. Why does that make a plant cell stiffer? The cell wall is made of a hard material. The cell walls in a tree trunk are what makes wood hard.

Section Review

Main Ideas: All living things are made of cells. These cells grow and reproduce. This chart shows how animal and plant cells differ.

Type of Cell	Has Cytoplasm	Has a Nucleus	Has a Cell Membrane	Has a Cell Wall
Plant	Yes	Yes	Yes	Yes
Animal	Yes	Yes	Yes	No

Questions: Answer in complete sentences.

1. What is a cell?
2. Name the three parts found in most cells.
3. Why are plant cells stiffer than animal cells?
4. What controls the activities of a cell?

CHAPTER REVIEW

Science Words: The clues in column B will help you unscramble the words in column A. Write your answers on a separate sheet of paper.

Column A	Column B
1. UCEDPORER	To make a copy like the first
2. POERCSCMIO	It makes small things look larger.
3. LLEC	A small unit of life
4. BREANMME	It covers all cells like a skin.
5. YTOPCLMAS	The living liquid inside a cell
6. EUCULNS	It is the captain of a cell.
7. VLINGONNI	The opposite of living

Questions: Answer in complete sentences.

1. What four things do all living things do?
2. Which of the following is not a living thing: a morning glory plant, a dog, or a car?
3. Give one reason why you know an icicle is not alive.
4. What are living things made of?
5. Name four parts of a plant cell.
6. Which of those parts is not found in an animal cell?
7. What does the nucleus of a cell do?
8. Why can't you see the cells in your body?
9. If you look at an onion skin under a microscope, you will see tiny boxes. What are these boxes called?
10. What is the skin-like covering that surrounds a cell?

CHAPTER 2

How Living Things Grow

2-1.

Where Do Cells Come From?

Tom has grown a lot this year. He is taller and heavier. His feet have also grown. Now he must have new shoes. What is happening inside Tom's body to make him grow?

When you finish this section, you should be able to:

- ☐ **A.** Describe how living things grow.
- ☐ **B.** Describe how an animal cell reproduces.
- ☐ **C.** Explain why things that are still growing need extra food.

To understand how living things grow, pretend you are building a tower of small blocks. You want to make the tower larger. What will you do? Maybe you will add more blocks.

Now, think of the cells in your body as little blocks. As you grow, the number of cells in your body also grows. When you are growing rapidly, thousands of new cells are added to your body each day. Do you know where these cells come from? They come from other cells.

Cell division: The way a cell makes two new cells.

In Chapter 1, you learned that living things can make copies of themselves. The single cells in your body can reproduce, too. A single cell reproduces by dividing into two cells. This is called **cell division** (dih-**vih**-zhun).

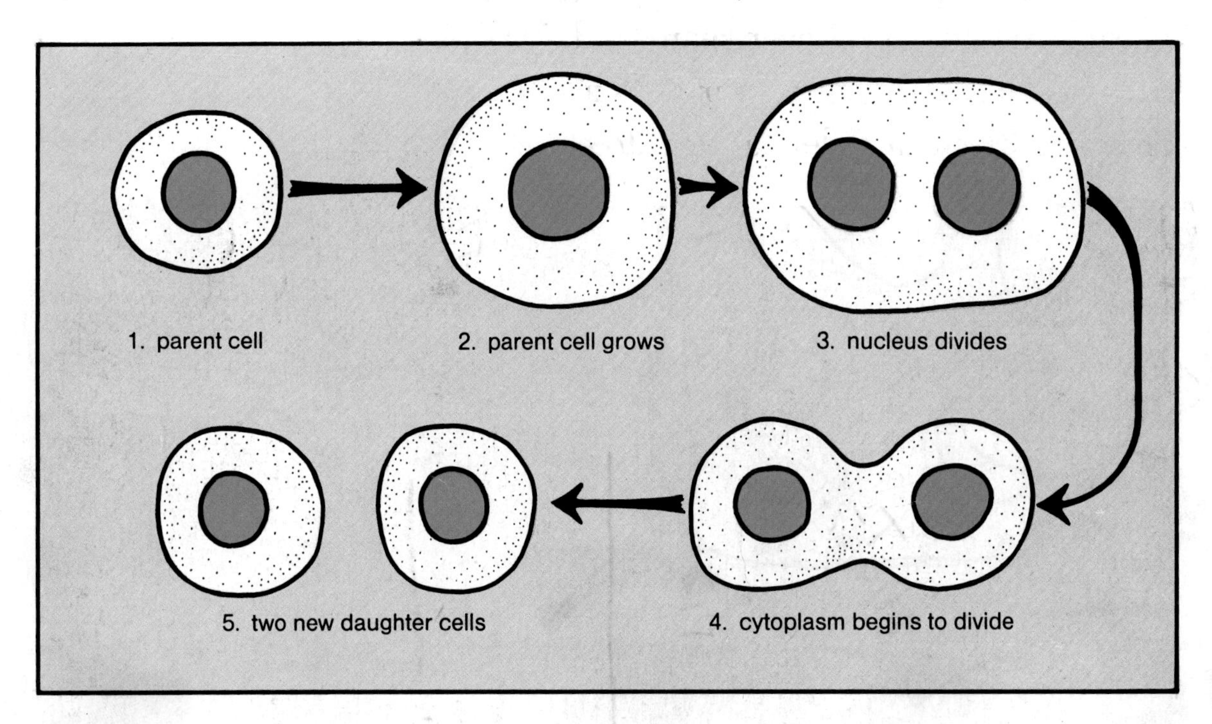

This is how an animal cell reproduces. First, the cell takes in extra food. It becomes larger. Next, the material of the nucleus begins to divide. When this division is complete, the cytoplasm begins to divide. Finally, the two parts of the parent cell pull apart. It becomes two new cells. These new cells are called daughter cells. Each daughter cell has a cell membrane, cytoplasm, and a nucleus, just like its parent cell. Each new cell can do what its parent cell could do. It can also make copies of itself. Plant cells can also reproduce in this way.

Has anyone ever told you that you must have a hole in your stomach because you eat so much? Do you get hungry when you are playing? If you are active, you need extra food for energy. If you are growing, you need even more food. Do you know why? It is because your cells need food. They need food in order to reproduce.

Humans usually stop growing at about age twenty. But their cells keep growing. New cells take the place of old cells. Do you know why a cut on your arm heals? New cells take the place of the cells that have been hurt.

Other types of living things grow at different rates. A baby kangaroo is less than 2 1/2 centimeters (1 inch) long when it is born. For the first six months of its life, it lives in a pouch (**powch**), or pocket, on its mother's stomach. By the time it is two or three years old, the kangaroo is fully grown.

Some plants, such as marigolds, become adult plants in a short time. Marigold seeds are planted in the spring. They grow very quickly. During the summer they bloom. At the end of summer, the plants die.

Section Review

Main Ideas: As living things grow, the number of cells in their bodies increases. Cells increase by making copies of themselves. To grow and to reproduce, cells need food.

Questions: Answer in complete sentences.

1. What happens to the number of cells in your body when you are growing?
2. Where do cells come from?
3. Describe how an animal cell reproduces.
4. Why do your cells need food?

Beatrice Sweeney

Dr. Beatrice Sweeney teaches biology at the University of California at Santa Barbara and at the College of Creative Studies, which is a school for gifted students.

Dr. Sweeney has always enjoyed working with plants. As a child, she drew and photographed plants. At first, she thought she wanted to be an artist. But when she was in college, she developed a scientific interest in plants.

The study of plants has taken Dr. Sweeney to many parts of the world. In 1960, she studied algae in northern Australia. She has also made scientific trips to New Guinea and Southeast Asia.

2-2.

Single Cells and How They Divide

Multicellular: Made of more than one cell.

Jan and Martin found a large frog at the pond. "Boy, is this frog big!" said Martin. "Just think," said Jan, "last year it was a single cell." What did she mean?

Frogs, like humans, are made of many cells. They are called **multicellular** (mul-tee-**sel**-yoo-ler) animals. *Multicellular* means made up of more than one cell. Do multicellular animals begin as one cell?

When you finish this section, you should be able to:

☐ **A.** Explain what an egg is.

☐ **B.** Describe how animals and plants develop from fertilized eggs.

Look at these pictures. All of these are eggs. Many living things develop from eggs. Chickens come from eggs. Can you find the chicken egg here? Frogs come from eggs. Do you know any other things that come from eggs?

An egg is a single cell. An egg has a nucleus, a cell membrane, and cytoplasm.

You know how one cell can become many cells. Do you know how an egg can become a frog? An egg from the female is fertilized (**fer**-tih-lized) by the male. The fertilized egg then divides into two cells. The two cells divide into four cells, and so on. Finally, a whole animal develops.

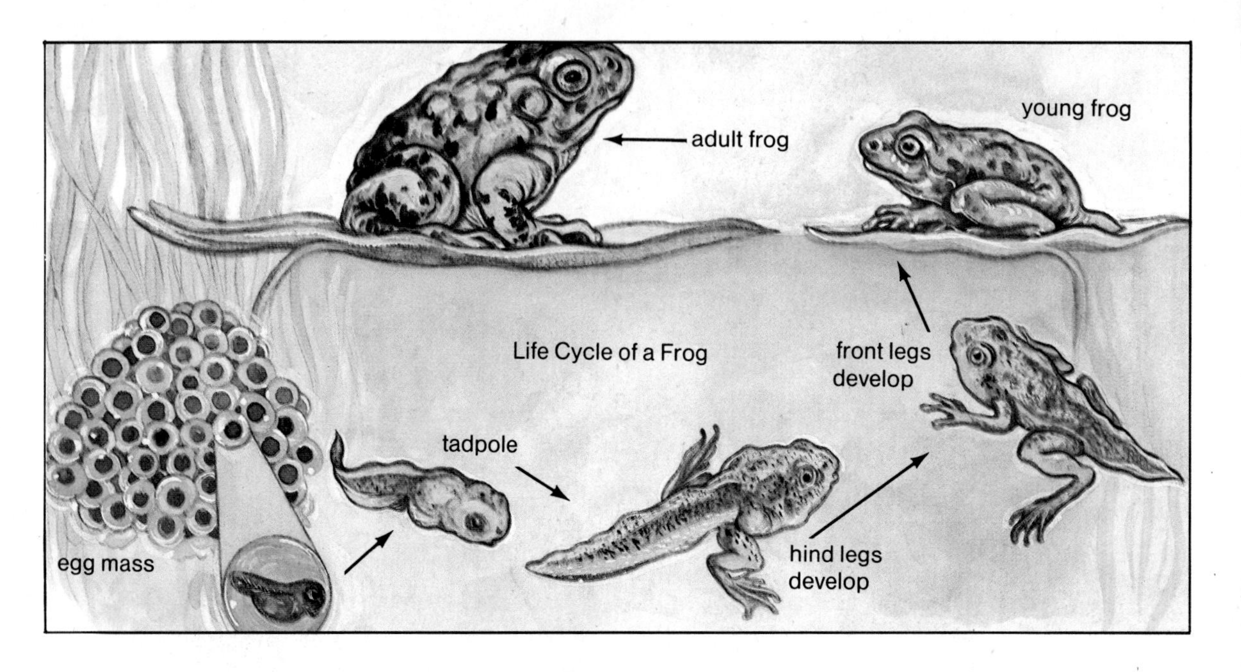

Mass: A collection or a group.

The eggs of a frog are small and soft. **Masses** of them are laid together in a kind of jelly. Young frogs then develop from the eggs in stages. These stages are shown above.

It only takes a few weeks for a frog egg to hatch. A wiggling tadpole comes out of the egg. As the tadpole grows, its body changes. Do you know what happens to the tadpole? The tadpole becomes a frog.

Embryo: A living thing that has not developed; a living thing in its early stage of growth.

You know that many plants produce seeds. Seeds contain **embryo** (**em**-bree-oh) plants. An *embryo* is an undeveloped living thing. If you plant a seed, the embryo plant inside it will begin to grow.

Where does the embryo plant come from? It comes from a fertilized egg. Like animals, plants produce egg cells. The egg cells of a seed plant develop within its flowers.

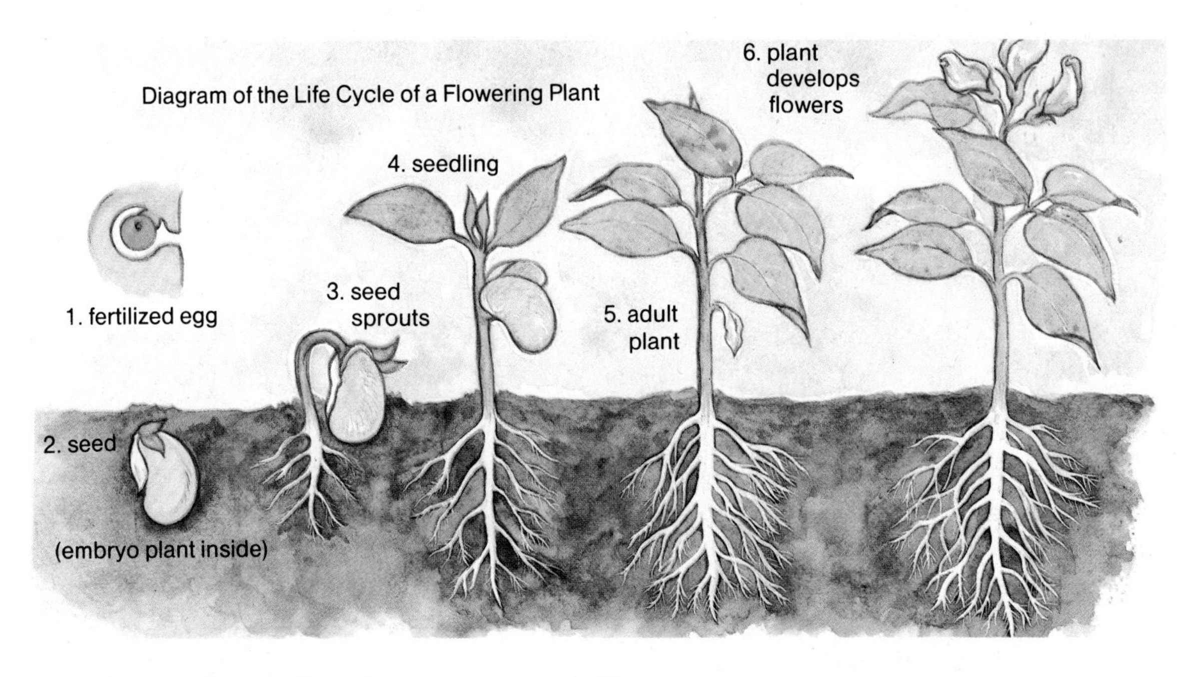

Young plants develop in stages. These stages are shown in the picture above.

ACTIVITY

Where are the egg cells and the seeds of flowers?

A. Gather these materials: young flower, seed pod, knife, and hand lens.

B. Cut the young flower in half with the knife.

C. Look at the flower with the hand lens.

1. Where are the eggs?

2. What do they look like?

D. Cut the seed pod in half.

3. Can you find the seeds? Where are they?

4. What is the difference between a young flower and a seed pod?

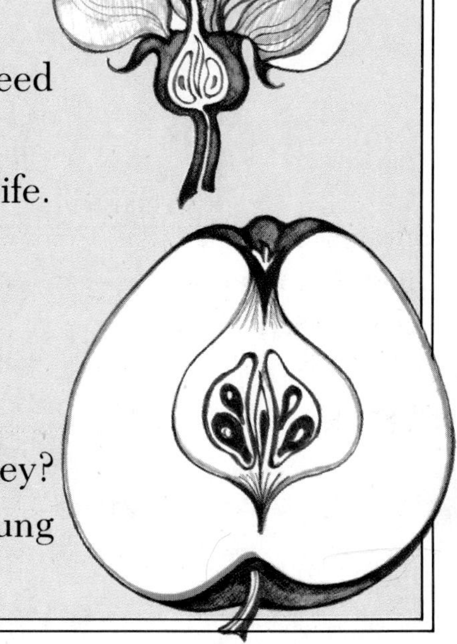

After a seed falls to the ground, what happens? When does a plant develop flowers? What will the flowers produce? Then what will happen? The picture on page 29 will help you answer all these questions.

Section Review

Main Ideas: An egg is a single cell. Most living things develop from fertilized eggs. This chart shows the stages of development of a frog and of a plant.

Type of Living Thing	Stages of Development				
Frog	Female produces an egg	Fertilized egg becomes tadpole	Tadpole develops legs	Tadpole becomes a frog	Frog grows up
Plant	Egg forms inside a flower	Seed develops from fertilized egg	Seed begins to grow	Seed becomes a plant	Plant develops a flower

Questions: Answer in complete sentences.

1. Name three things that come from eggs.
2. What is an embryo?
3. How does a single cell develop into a living thing that has many cells?
4. Name the stages in the life cycle of a frog.

2-3.

Special Cells

Can you name some parts of your body? Your answers are probably the head, neck, legs, arms, feet, and hands. All parts are important. We use different parts to do different things. For example, we use our hands to write, to eat, to play ball, and to do many other things. Often, we use several parts at the same time. When you finish this section, you should be able to:

- ☐ **A.** Explain how your body is made.
- ☐ **B.** Describe some different cells in your body.
- ☐ **C.** Describe some different cells in a plant.

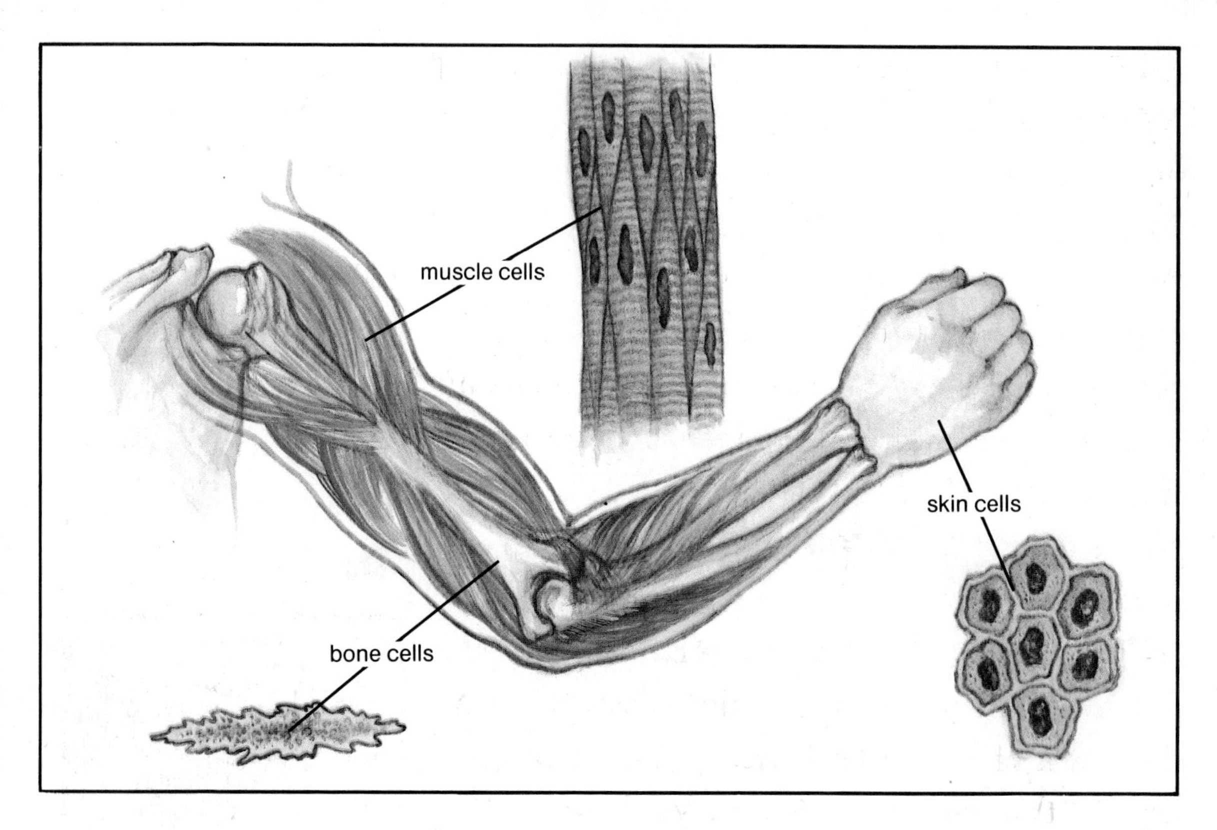

It is easy to name the parts of the body. Now observe your body more closely. Do you know what each part is made of?

Let's look at your arm. You can see that it is covered by skin. Now feel your arm. Beneath the skin is muscle (**muss**-ul) and bone. These parts work together. They help your arm carry out its activities.

Suppose you could study pieces of muscle, skin, and bone under a microscope. You would find that each piece is made of cells. But the cells are not the same in each piece. Muscles do not do the same work as skin. So, muscle cells are different from skin cells. The above picture shows three different kinds of cells.

In the human body, most different types of cells have a nucleus, cytoplasm, and a cell membrane. But each type is shaped to do different work. Cells that are like each other work together to do their special jobs.

Plants also have different kinds of cells. Some of the cells of a green plant have little green bits in their cytoplasm. These green bits are called **chloroplasts** (**klore**-uh-plasts). They give green plants their color. *Chloroplasts* also help green plants make food.

Chloroplasts: The green bodies in cytoplasm; bodies that give plants their color and help them make food.

The trees in the picture are flowering plants. The leaves have cells with chloroplasts. Cells in the trunk of a tree have heavy walls that help support the tree trunk. Plant roots have special cells that grow root hairs. These root hairs help the plant take in water.

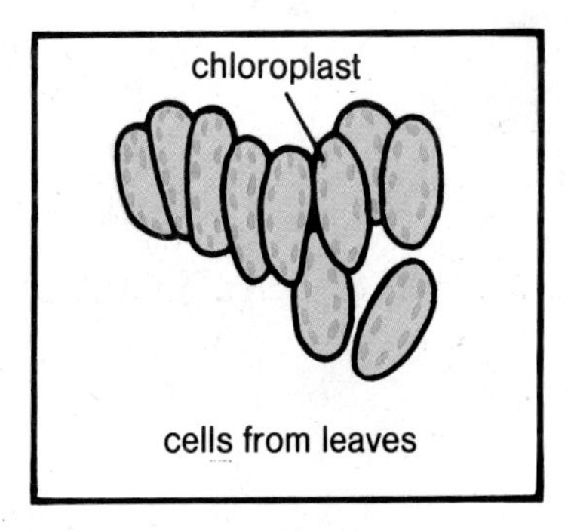

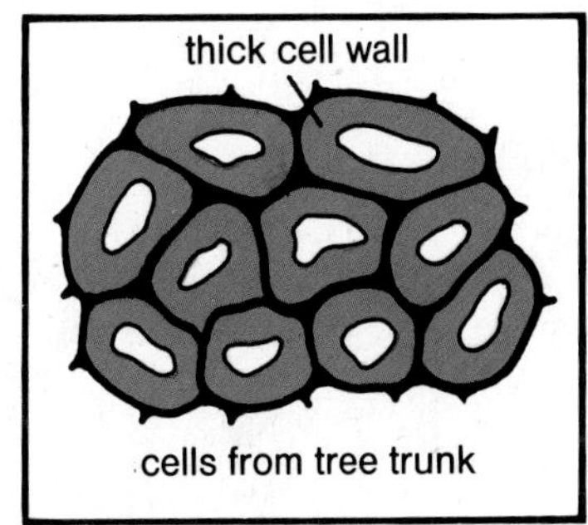

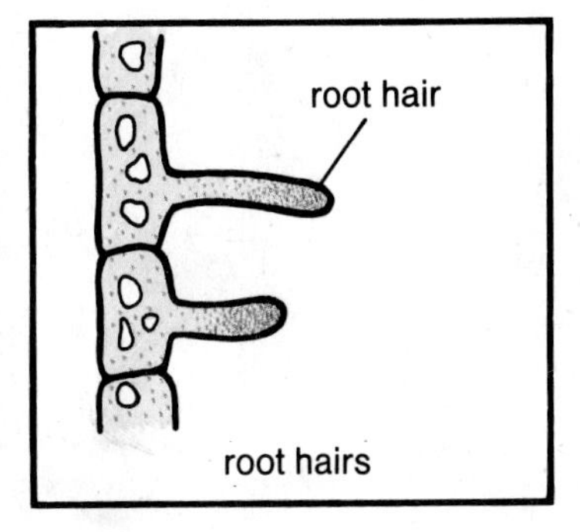

Organism: A complete living thing; it can have one cell or many cells.

You have learned that a multicellular living thing develops from a fertilized egg cell. You may wonder how different kinds of cells can develop from one kind of cell. All cells have special growing instructions in their nuclei (**noo**-klee-eye). Remember, a nucleus is like a cell captain. The nucleus of a cell controls what all new cells will be like. When a fertilized egg divides, the nucleus decides what the new **organism** (**or**-gan-iz-um) will look like. An *organism* is a complete living thing. Look at the pictures. The frog came from one of the eggs shown at the left.

Section Review

Main Ideas: Most multicellular animals and plants are made of different kinds of cells. Each kind of cell does its own special kind of work. The nucleus of a cell controls how a living thing grows and develops. Some types of cells are listed in this chart.

Type of Living Thing	Some Types of Cells
Human Body	Skin Muscle Bone
Tree	Leaf Trunk Root

Questions: Answer in complete sentences.

1. Why are muscle cells different from skin cells?
2. Why do roots have root hairs?
3. What part of the egg decides what the new organism will be like?
4. What are the little green bits in a cell of a plant?
5. Name two parts of the arm that help it carry out its activities.
6. Name four parts of the human body.

CHAPTER REVIEW

Science Words: Write the sentences below on paper. Fill in the blanks with the correct words from the list.

chloroplasts	**tadpole**
organism	**root hairs**
division	**embryo**

1. Cell ______ is a way cells reproduce.
2. A ______ is a developing frog.
3. Inside every seed is a tiny ______ plant.
4. ______ give green plants their color.
5. ______ help plants take in water.
6. An ______ is a complete living thing.

Questions: Answer in complete sentences.

1. What does multicellular mean?
2. What are the little bits of green in the cell of a plant leaf?
3. Name three types of cells found in the human body.
4. Name three special kinds of plant cells.
5. How long does it take for a frog egg to hatch?
6. What are the three main stages in the development of a frog?
7. Where does a baby kangaroo live during the first six months of its life?
8. Why does a cut on your arm heal?
9. What part of a cell controls how a living thing grows?

CHAPTER 3

SIMPLE LIVING THINGS

What is the green in this picture? It is found in many streams and ponds. You will learn about this and other one-celled living things in this chapter.

3-1. One-celled Organisms

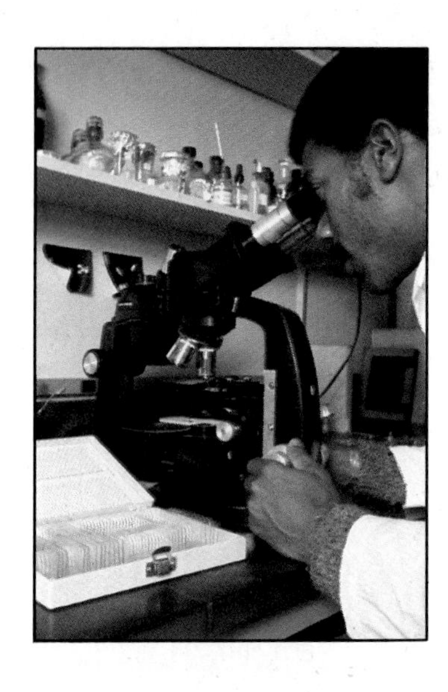

When you finish this section, you should be able to:

- ☐ **A.** Describe some of the smallest living things.
- ☐ **B.** Explain how these living things get food.
- ☐ **C.** Explain how a single cell can be a complete living thing.

Look at the pictures below. The odd-looking things are "wee beasties." They were found 300 years ago by a man named Anton van Leeuwenhoek (**lay**-vun-hoohk). Leeuwenhoek made microscopes. One day, he looked at a drop of water under one of his microscopes. He was very surprised at what he saw. Tiny creatures were wiggling in the water. Some were grouped together. They had many different shapes. Look at these shapes in the picture.

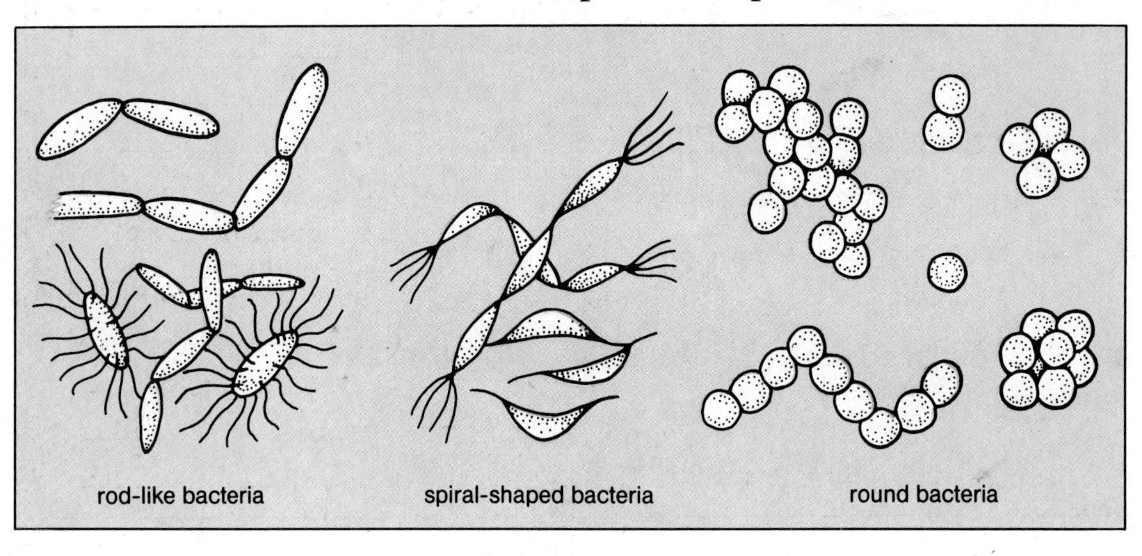

Today these "wee beasties" are called **bacteria** (bak-**teer**-ee-uh). *Bacteria* have only one cell. They are so small that thousands could fit onto the head of a pin. Yet bacteria are complete living things.

Bacteria: Very small one-celled living things.

How can one cell be a whole living thing? It can do what a larger living thing can do. It uses food. It grows and changes. It moves. It reproduces. It does these things by itself as one cell.

It is hard to imagine how many bacteria live on earth. They live everywhere and on everything. There are more bacteria than there are blades of grass. Most bacteria get their food from the things they live on.

ACTIVITY

Can you make charts that list foods made by bacteria and diseases caused by bacteria?

A. Gather these materials: poster paper, ruler, felt-tipped pen, and science books.

B. Copy these charts on the poster paper.

C. Read the science books to find out more about bacteria.

1. What diseases are caused by bacteria?
2. What foods are made with the help of bacteria?

D. Record your findings on the charts.

Foods Made with Help of Bacteria

Diseases Caused by Bacteria

Some bacteria live in the human body. Most do not harm the body. But some cause disease. Other bacteria live on food. Some of them cause food to spoil. Others help us make food. For example, bacteria help us make buttermilk and sour cream.

Protozoan: A one-celled animal-like living thing.

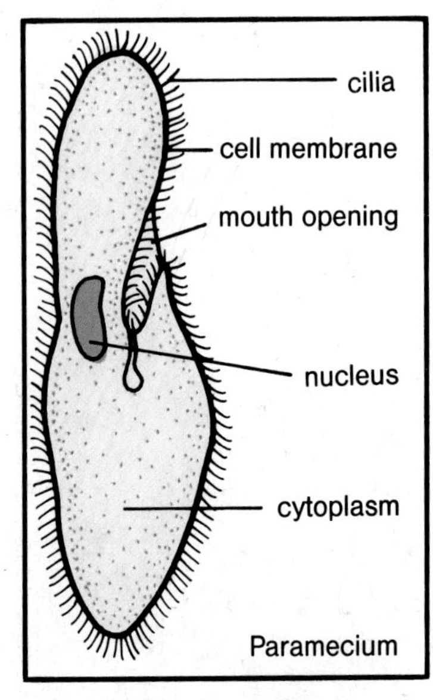

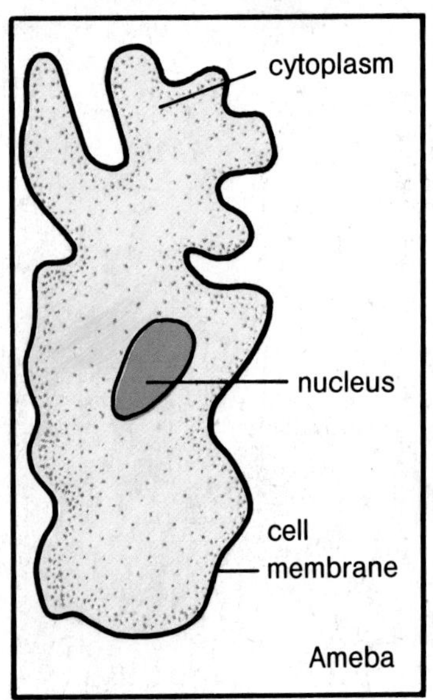

You would need a powerful microscope to see bacteria. But you could use a classroom microscope to see the small living things shown in the pictures in the margin. Leeuwenhoek found some of these things in pond water. Each one belongs to a group of animal-like creatures called **protozoans** (proh-toh-**zoh**-uhns). Like bacteria, *protozoans* have only one cell. Yet, they are complete living things.

One type of protozoan is a paramecium (pair-uh-**mee**-see-um). This slipper-shaped creature is circled by little moving hairs. A paramecium moves through water by waving these little hairs. It also uses these hairs to get food into its mouth. This mouth is really just a hole in the side of the paramecium. Can you guess what it eats? It eats bacteria, other protozoans, and tiny plant-like creatures.

Another kind of protozoan is the ameba (uh-**mee**-buh). An ameba has no real shape. As it moves, its shape changes. First, it pushes a piece of itself forward. Then, it drags the rest of itself along. As you can imagine, the ameba moves very slowly.

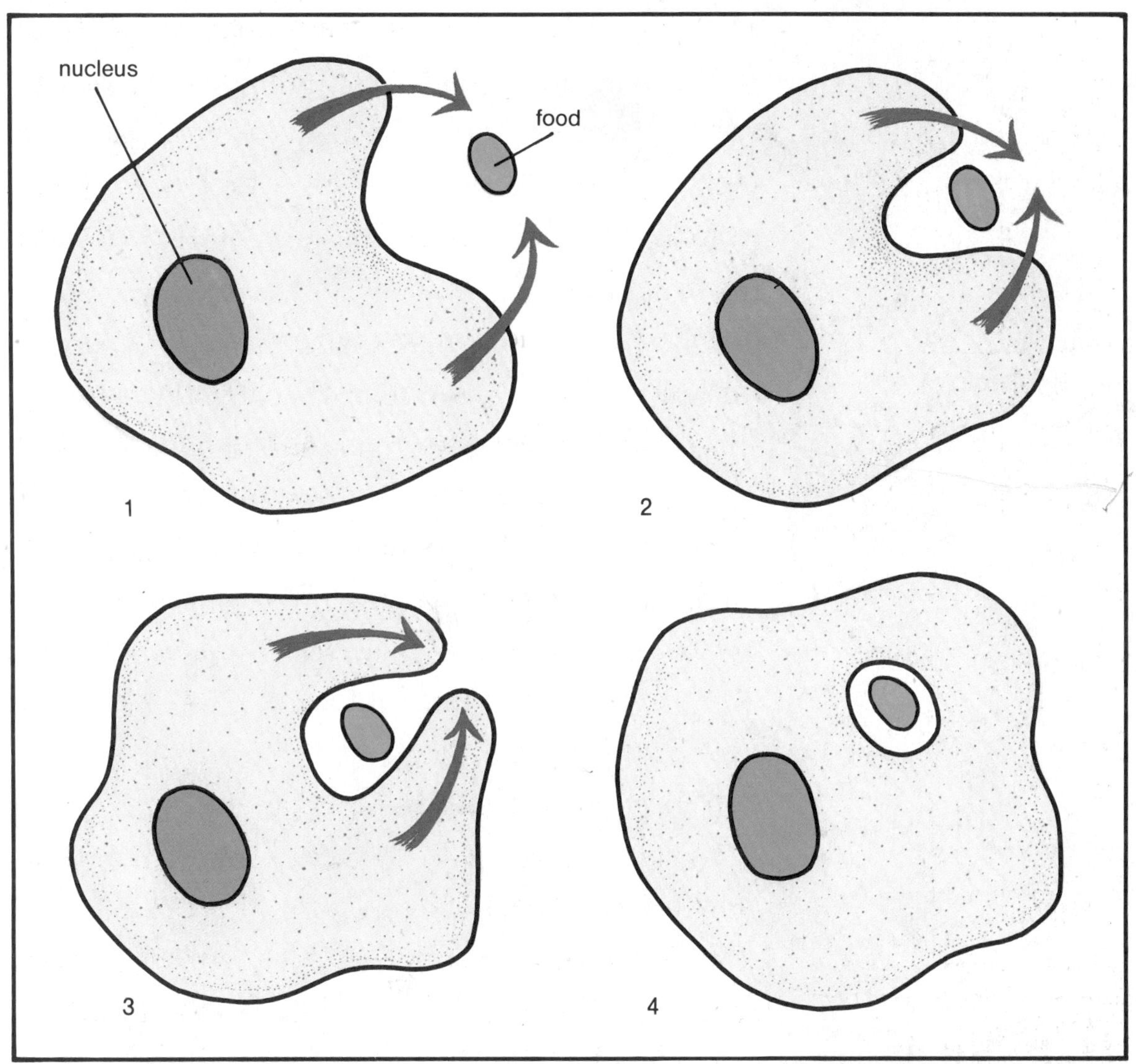

As shown in the picture above, an ameba changes its shape to catch food.

A very unusual protozoan is the euglena (yoo-**glee**-nuh). It is bright green. A euglena moves by using its thread-like whip. The whip is also used to move food into the euglena's mouth. Euglenas eat bacteria, other protozoans, and plant-like living things. But they can also make food. Notice that a euglena has chloroplasts. Do

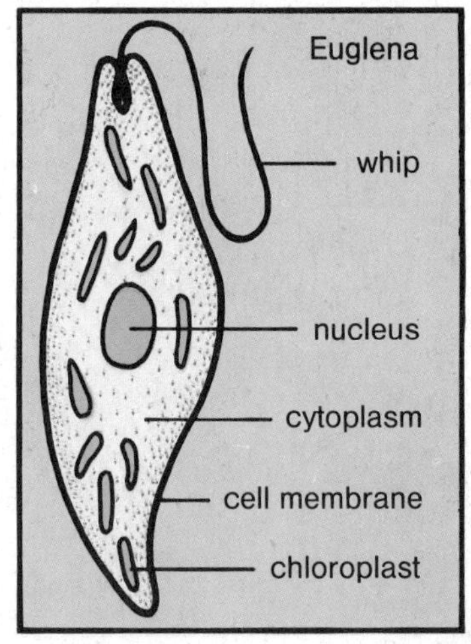

you remember what chloroplasts do? What other living things have chloroplasts? In the next section, you will find out how living things use chloroplasts to make food.

Whales are huge organisms. Bacteria and protozoans are tiny organisms. What is the main difference between whales and these tiny organisms? It is the number of cells they have. Bacteria and protozoans are one-celled organisms. One tiny cell carries out all the activities for these organisms. Whales have many cells. In a whale, billions and billions of cells work together to keep the whale alive. Remember, different kinds of cells do different kinds of work.

Section Review

Main Ideas: The smallest living things have only one cell. Yet, each one is a complete organism. Bacteria and protozoans are one-celled organisms. These tiny organisms are found everywhere.

Questions: Answer in complete sentences.

1. Where are bacteria found?
2. Where do most bacteria find their food?
3. What is a protozoan?
4. Describe a paramecium.
5. What is the main difference between whales and bacteria?

3-2. Other One-celled Organisms

In the spring, the water in this pond was clear. You could look through it and watch the tadpoles. Now it is summer. A thick, green scum covers one part of the pond. You cannot see through this scum. What is this scum? Here's a hint. It has chloroplasts.

When you finish this section, you should be able to:

- ☐ **A.** Name the living thing that causes pond scum.
- ☐ **B.** Tell how chloroplasts help living things make their own food.

This is how a small piece of pond scum looks under a microscope. Pond scum is really chains and chains of plant-like cells. The cells have chloroplasts, which are shaped like coils.

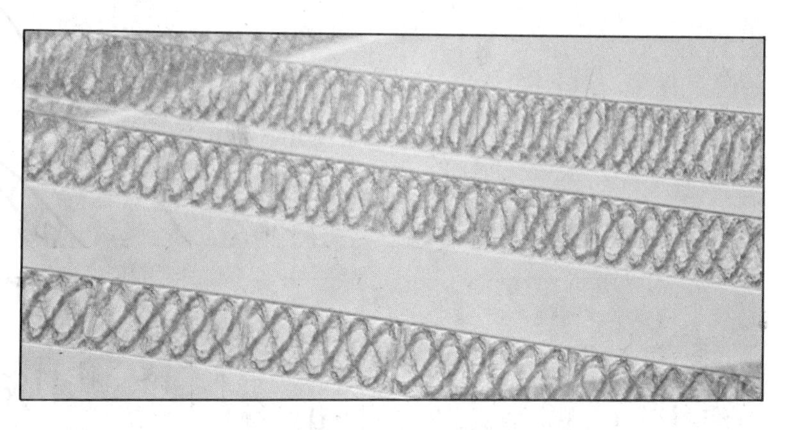

Algae: Plant-like organisms with chloroplasts; they can make their own food.

Is pond scum a kind of plant? No, not quite. Pond scum is a kind of **algae** (**al**-jee). Like plants, *algae* have chloroplasts. But algae do not have stems, leaves, and roots as true plants do.

Many kinds of algae live in the waters of the world. Still others live on rocks, on trees, or on soil. Some algae are one-celled organisms. Often, they float by themselves in water. Some,

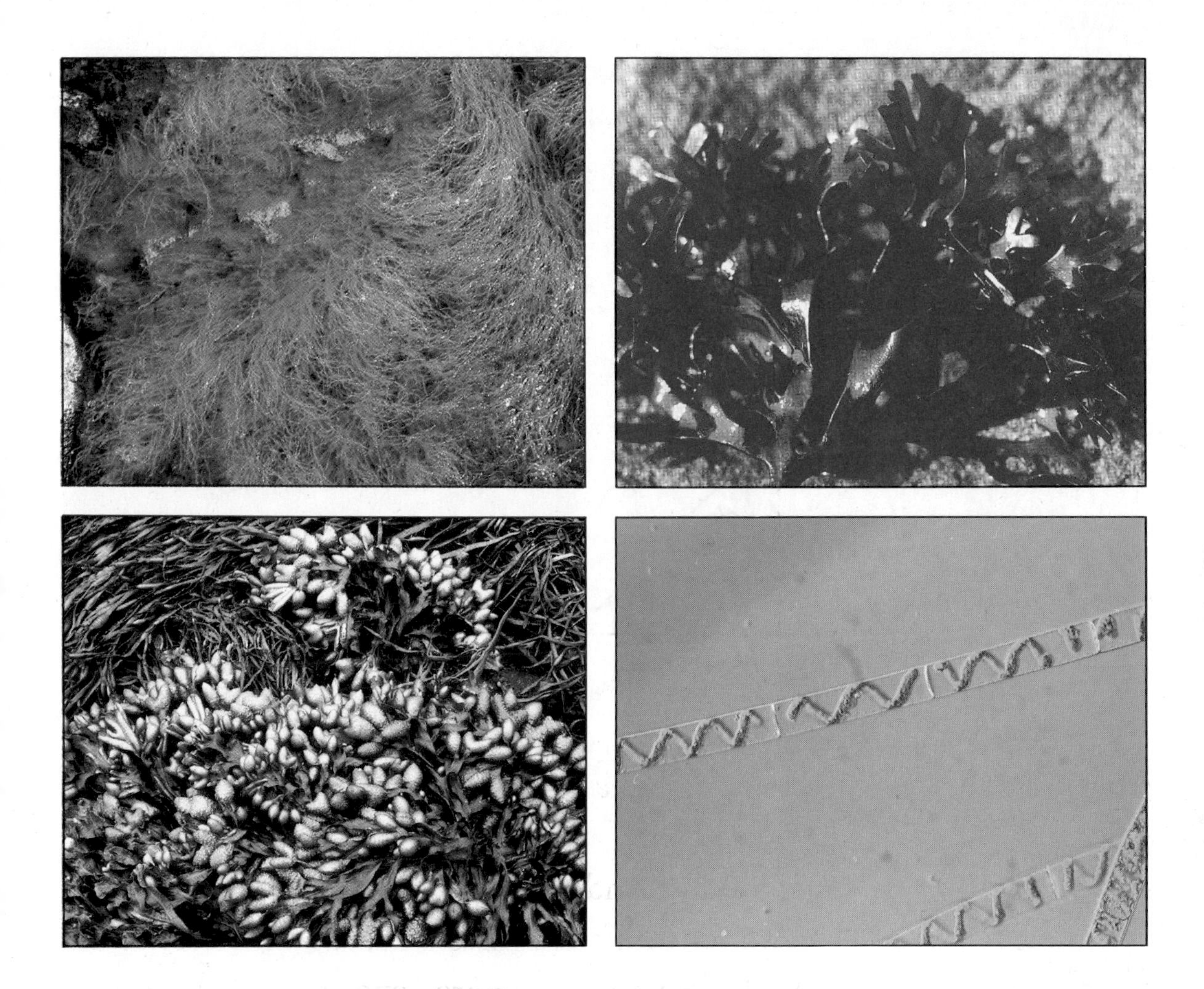

like pond scum, live together in chains. Others live together in large groups and have root-like ends. These help them hold on to rocks or soil. Some of these algae become very large.

Most algae are green. However, they can also be other colors. Examples of green, brown, blue-green, and red algae are shown above.

Many of these algae live in the ocean. Some have tiny "balloons" on the ends. These help the algae float in the water. Some other algae can grow as long as a classroom.

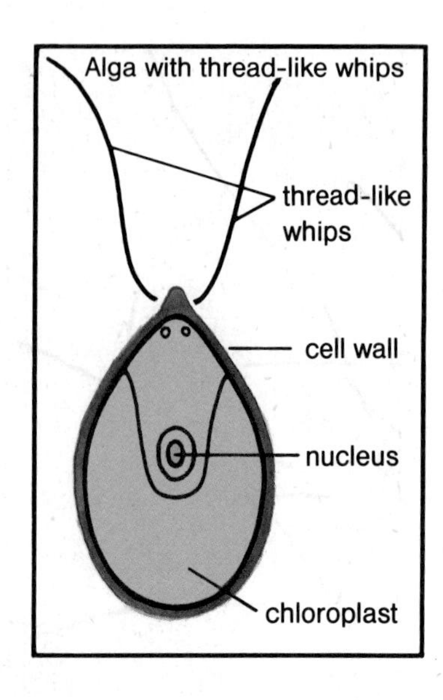

Some one-celled algae can move themselves around. These tiny algae have thread-like whips. They go through the water by moving these whips. Do you know another one-celled organism that moves this way?

When animals are hungry, they eat other organisms. When most protozoans are hungry, they also eat other organisms. Most plants, however, do not need to eat other organisms. Most plants can make their own food. Algae can make their own food, too. Algae make their food out of non-living things.

Look at the picture below as you read this paragraph. To make food, a cell needs energy. The chloroplasts inside a cell trap energy from the sun. Next, water is needed. Water enters a cell

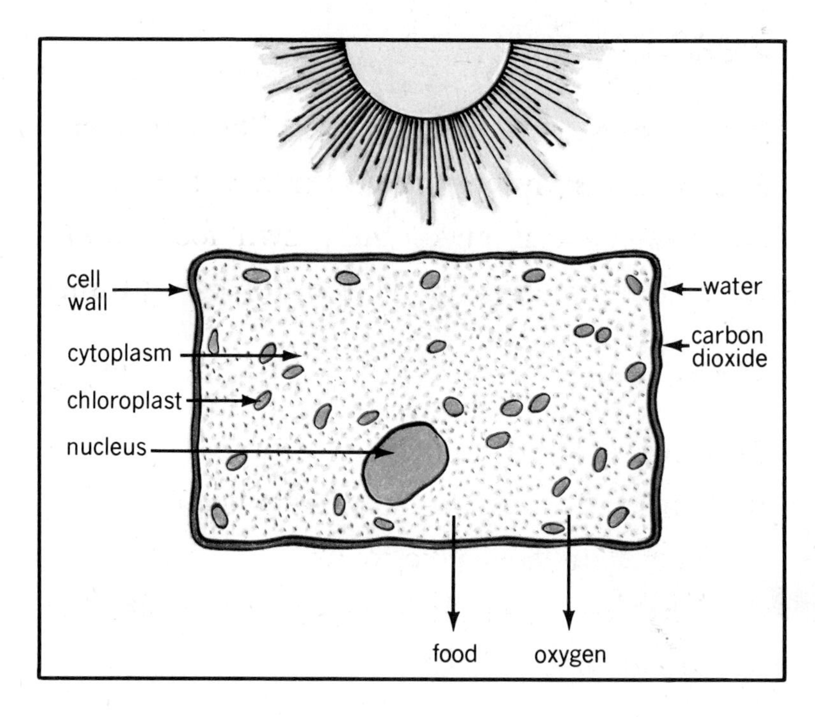

through the cell wall. Finally, a type of gas, called carbon dioxide (**kahr**-bun dy-**ahk**-side), is needed. This gas enters the cell through the cell wall. Using energy from the sun, the cell turns the water and the gas into food and oxygen (**ahk**-sih-jun). The food can then be used by the cell. The oxygen can be used by other living things.

Green plants are sometimes called food factories. This is because they can make food out of non-living things. Algae are also little food factories. Algae use the food they make to live and grow. But sometimes protozoans, fish, and other creatures eat algae. When this happens, who uses the food made by these little food factories?

Section Review

Main Ideas: Algae are plant-like organisms. Some have only one cell. Others are groups of cells. Algae can make their own food. Sometimes they become food for other living things.

Questions: Answer in complete sentences.

1. What is pond scum?
2. What color are the simplest algae?
3. How do algae make food?
4. Why are green plants sometimes called food factories?

3-3. Simple Food Users

How did the holes get into this bread? What made the edge of this bread a dirty gray?

Tiny plant-like organisms did this. One type of organism is used in bread making. The other type spoils food. Can you name these two organisms?

When you finish this section, you should be able to:

- ☐ **A.** Describe the plant-like organism used in making bread.
- ☐ **B.** Describe a plant-like organism that spoils food.
- ☐ **C.** Explain how these organisms live.

Have you ever tried to make bread? This is what you would use: flour, milk, butter, salt, sugar, and **yeast** (**yeest**). Do you know which of these is a living thing? The answer is *yeast*. Yeast is a one-celled organism. You can buy yeast in packages. A single yeast cell is too small to see without a microscope. But you can see the work of a group of yeast cells.

Yeast: A one-celled organism used in bread making.

Yeast needs sugar to live. When you add yeast to bread dough, it uses the sugar in the dough for food. The yeast begins to grow and reproduce. As it does, it gives off a gas. This gas cannot get out of the dough. It collects in bubbles. This makes the dough puff up.

Bakers always put bread dough in a warm place. The warmth helps the dough to puff up more quickly. Notice the little bubbles in the dough.

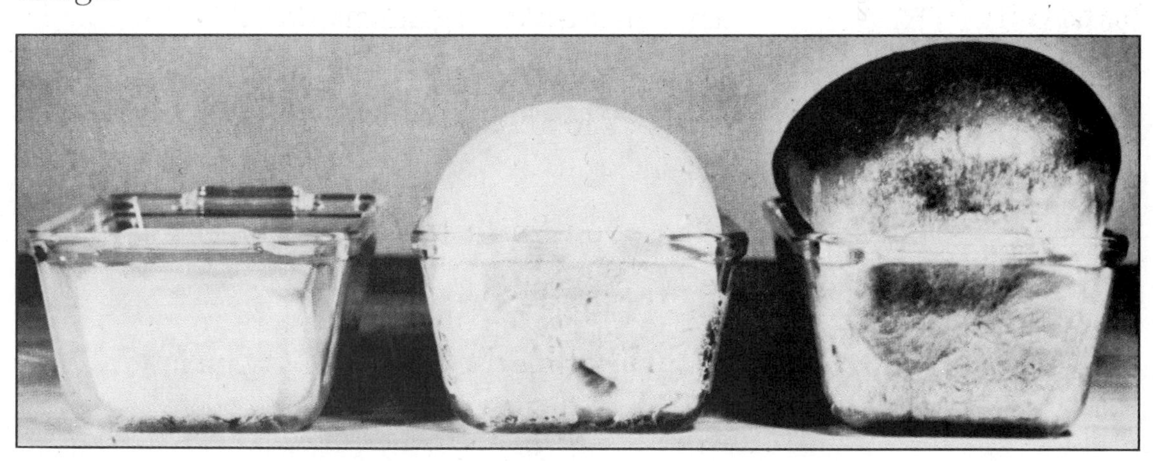

After bread dough is baked, there are little holes in it. What caused the little holes? It was the gas that the yeast produced.

ACTIVITY

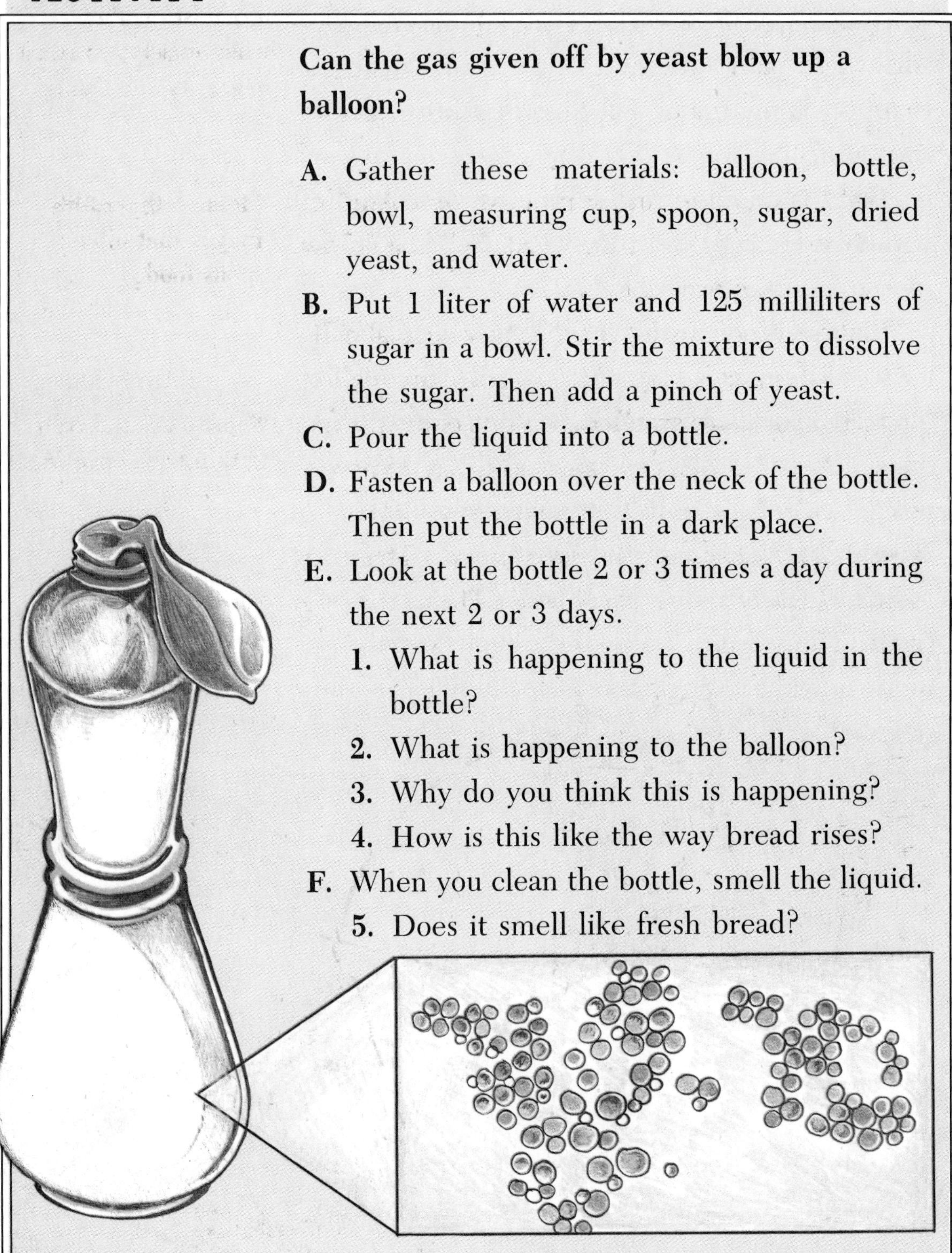

Can the gas given off by yeast blow up a balloon?

A. Gather these materials: balloon, bottle, bowl, measuring cup, spoon, sugar, dried yeast, and water.

B. Put 1 liter of water and 125 milliliters of sugar in a bowl. Stir the mixture to dissolve the sugar. Then add a pinch of yeast.

C. Pour the liquid into a bottle.

D. Fasten a balloon over the neck of the bottle. Then put the bottle in a dark place.

E. Look at the bottle 2 or 3 times a day during the next 2 or 3 days.

1. What is happening to the liquid in the bottle?
2. What is happening to the balloon?
3. Why do you think this is happening?
4. How is this like the way bread rises?

F. When you clean the bottle, smell the liquid.

5. Does it smell like fresh bread?

Yeast is a kind of **fungus** (**fung**-us). A *fungus* is a plant-like organism. It does not have chloroplasts. It cannot make its own food. A fungus lives on things that are alive or that were alive at one time.

Fungus: A plant-like organism without chloroplasts; it cannot make its own food.

Mold is another kind of fungus. One kind of *mold* grows on bread. Like yeast, mold can grow and reproduce very quickly.

Mold: A thread-like fungus that often spoils food.

Mold reproduces by making tiny special cells with a hard covering. These cells are called **spores**. *Spores* are smaller than bits of dust. Like dust, they can float through the air. When a mold spore falls onto bread, it grows. First, it sends little threads into the bread. Then, it sends threads above the bread. These threads make more spores.

Spores: Special cells with a hard covering.

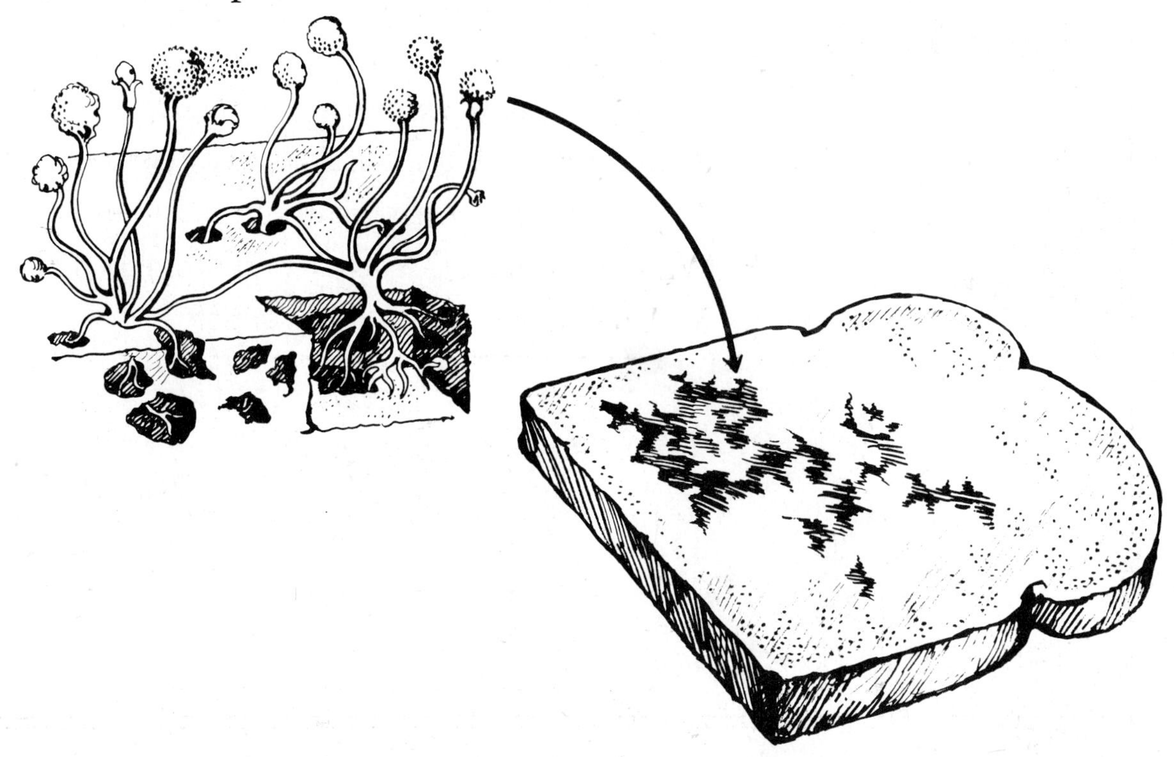

Bread mold makes bread taste bad. Another mold spoils oranges, grapefruit, and lemons. But some molds give food flavor. Have you ever eaten blue cheese? Mold is used in making blue cheese.

Section Review

Main Ideas: A fungus cannot make its own food. Therefore, a fungus must live on something that is alive or that was alive at one time. Some of these organisms are helpful. Some are harmful.

Questions: Answer in complete sentences.

1. Name a type of organism used in bread making.
2. What is a fungus?
3. Name two kinds of fungi.
4. Why can't a fungus make food?
5. Name an organism that can spoil food.

CHAPTER REVIEW

Science Words: List the letters **a** through **f** on paper. Write the correct word from the list next to each letter.

yeast	**mold**
algae	**euglena**
paramecium	**spore**

John put a drop of pond water under his microscope. He saw two protozoans, a slipper-shaped ___a___ and a bright green ___b___. He also saw a chain of green ___c___.

Next, John looked at an old slice of bread. The edge of the bread had spots of gray ___d___. A ___e___ from this plant-like organism had dropped into a hole in the bread. John knew that the gas that the ___f___ produced had caused the holes in the bread.

Questions: Answer in complete sentences.

1. How can one cell be a complete living thing?
2. What do protozoans eat?
3. What color are most algae?
4. Name and describe a one-celled organism that moves by using its thread-like whip.
5. Where do fungi get food?
6. What are spores?
7. How are whales different from protozoans?

INVESTIGATING

How much does a plant grow in a week? In 4 weeks?

A. Gather these materials: clay pot, metric ruler, soil, plastic bag, water, 4–5 beans, paper, and pencil.

B. Put soil in the pot and water it.

C. Plant the beans in the soil.

D. Put a plastic bag over the pot.

E. Set the pot in a warm place.

F. When the beans sprout, remove the plastic bag. Take out all but 1 of the sprouts.

1. How long did it take for the beans to sprout?

G. Place the pot in a sunny window. Water the soil.

H. Copy this chart on a sheet of paper.

Time	Height (in cm)	Number of Leaves
1st week		
2nd week		
3rd week		
4th week		

I. Observe the pot each week. Measure how much the plant has grown. Record your findings.

2. How much did the plant grow between weeks 1 and 2? Between weeks 2 and 3? Between weeks 3 and 4?

J. Keep the plant growing as long as possible.

CAREERS

Botanist ▶

You have learned how plants reproduce and grow. A botanist is a person who studies plant life. Some botanists study only one group of plants, such as flowering plants. Other botanists study plant-like organisms, such as algae. People who are botanists have studied science in college.

◀ Lab Assistant

You have seen how much larger cells look through a microscope. Some lab assistants study bacteria in laboratories. By looking at bacteria under a microscope, they can help find ways to prevent disease. A lab assistant has studied science. Two or more years of college are needed.

ROCKS AND FOSSILS

UNIT 2

CHAPTER 4

ROCKS

4-1. Observing Rocks

This person is mountain climbing. Do you think it would be hard to climb to the top of these rocks? Have you ever walked or climbed in the mountains?

When you finish this section, you should be able to:

- ☐ **A.** Tell where rocks are found.
- ☐ **B.** Name some *properties* of rocks.
- ☐ **C.** Explain what a *mineral* is.

Rocks are found all over the earth. Let's pretend we are on a trip. First, we will go to a mountain. All mountains are made of rocks. The tops of some mountains are very high. The rocks here look very hard.

Now let's go to a flat area. The rocks here are in layers. The picture on the right shows the different layers. How are these rocks different from the rocks in the other picture?

Property: Something you can observe, such as color, shape, or feel.

At the beach, we find round and smooth rocks. What has happened to these rocks? What made them smooth?

Look at the rocks in the picture below. Some of them came from mountains. Others came from flat areas. And some were found on the beach.

The rocks are different. They have different **properties**. A *property* is something you can observe, such as color, shape, or feel. One example of a property is color. What colors do you see in these rocks? You can observe another property by touching a rock. Some rocks are smooth. Others are rough. A rock can also have stripes. The stripes look like layers. Rocks can also be shiny or dull. Can you name other properties of these rocks?

Pretend that you are holding the rock shown at the right in your hand. Look at it closely. Do you see tiny pieces of different colors? These are **minerals** (**min**-uh-rulz). All rocks are made of *minerals*. Minerals are materials found in the earth. Some rocks are made of only one mineral. Other rocks are made of more than one mineral.

One of the most common minerals is **quartz** (**kworts**). There are many types of *quartz*. Each type is a different color. Four types are shown below. Quartz is very beautiful. This is why it is often used in jewelry.

Some minerals can be eaten. One makes your food taste salty. It is called **halite** (**hayl**-ite). *Halite* is found in the earth. Can you name some other minerals?

Minerals: The materials that rocks are made of.

Quartz: A common mineral; can be different colors.

Halite: A mineral that tastes salty.

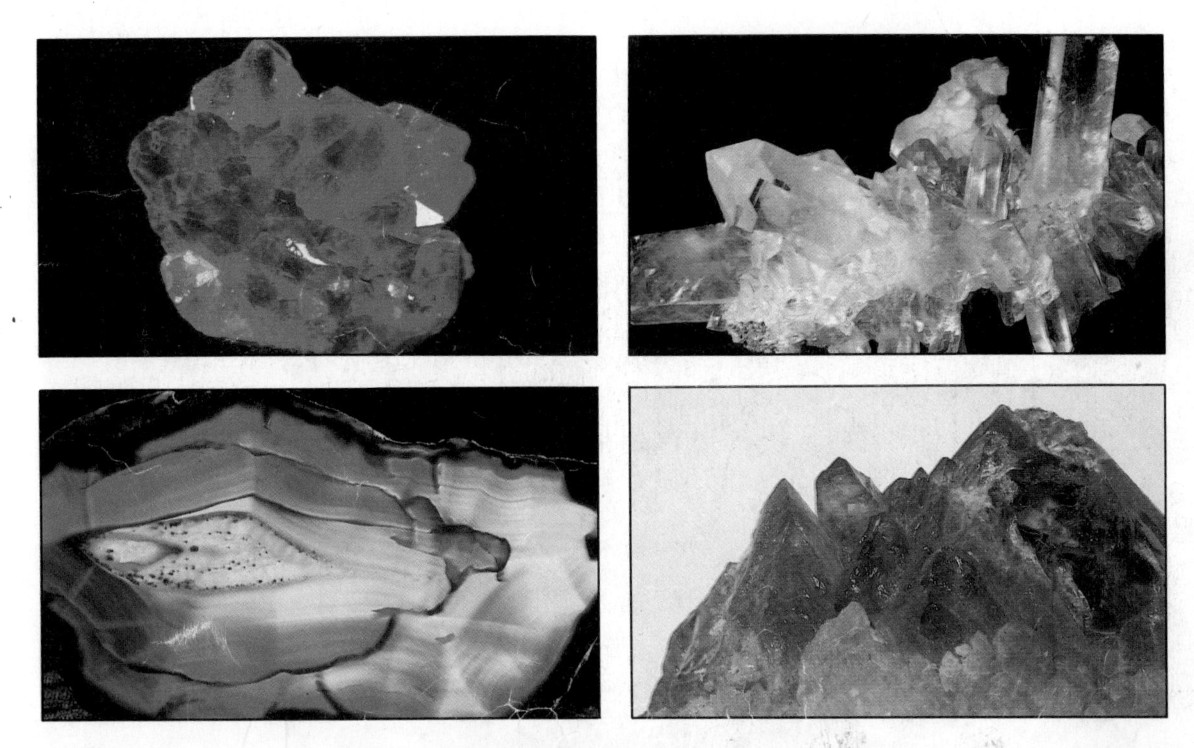

Activity

How are rocks different?

A. Gather these materials: 5 rocks, masking tape, and hand lens.

B. Using masking tape, label each rock: A, B, C, D, and E.

C. Answer these questions about each rock.

1. What color is the rock?
2. Are there tiny pieces in the rock? If so, what are their shapes and colors?
3. Is the rock rough or smooth?
4. Does the rock have stripes?
5. Is the rock shiny or dull?

Section Review

Main Ideas: Rocks are found everywhere. We can observe the properties of rocks. Rocks are made of minerals.

Questions: Answer in complete sentences.

1. Look at the two kinds of rocks in the picture. List two properties for each rock.
2. Where do you think you could find these rocks?
3. What is a mineral?
4. What mineral is often used in jewelry? Why?

4-2.

Using Rocks

Have you ever tried to skip a rock across water? Did it skip over the water and then sink? Most rocks sink in water. Look at this picture. This rock is floating. Do you know why it can float?

When you finish this section, you should be able to:

- ☐ **A.** Name some uses of rocks.
- ☐ **B.** Explain how rocks can be grouped by their hardness.

Rocks can be used in many ways. You decide how a rock can be used by its properties. Have you ever used a compass? Like a compass, the rock shown below can help you find direction. It has a property that most rocks do not have. When it is held by a string, it will turn and point north. Do you know why this happens? Because the rock acts like a magnet.

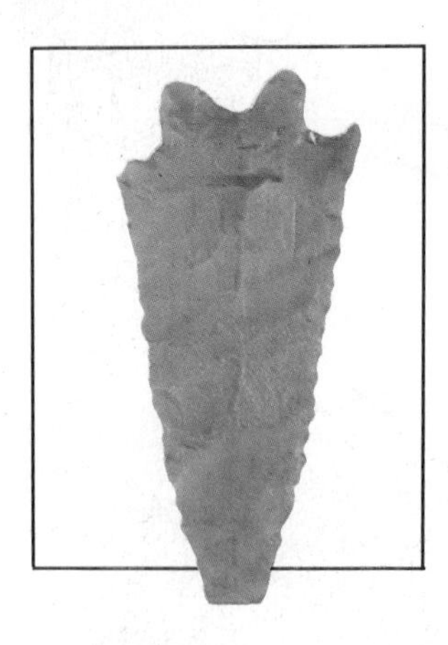

Let's look at some other uses for rocks. These rocks are very hard. They were used as arrowheads. Notice the pointed ends. Why would hard rocks be used as arrowheads?

Limestone: A soft rock.

This cave was carved out of **limestone** by water. *Limestone* is a soft rock. It is not as hard as the rocks that were used as arrowheads. Limestone is often used as a building material. Do you know other uses for limestone? It can also be used like a pencil or chalk. It is easy to rub limestone off paper and chalkboards.

This boy is scratching glass with a rock. You can see that the rock scratches the glass. Glass is hard. But the rock is harder. A scratch test can be used to test for **hardness**. *Hardness* is a property of rocks and minerals. Some rocks can be scratched with your fingernail. These rocks are not very hard. They are soft. To find the hardness of rocks, all you need is your fingernail, a penny, a knife, and a piece of glass.

Hardness: A property of rocks and minerals found by using the scratch test.

Diamond: The hardest mineral.

Hard rocks have different uses than soft rocks. One of the hardest minerals is **diamond** (**dy**-uh-mund). *Diamond* is so hard that it can be used to cut other rocks. Diamonds are also used in jewelry. Why do you think they are used for this purpose?

Look at the picture on the right. The bird was shaped by a sculptor. How do you think its hardness compares to the hardness of diamond?

ACTIVITY

Can you find the hardness of a rock or mineral?

A. Gather these materials: glass plate, penny, nail, masking tape, and 4 rocks or minerals.

B. Make a chart like this. Then, using masking tape, label each rock: A, B, C, and D.

Rock	Fingernail	Penny	Nail	Glass
A				
B				
C				
D				

C. Try to scratch each rock with your fingernail, the penny, and the nail.

D. Try to scratch the glass with each rock.

1. Which rocks were you able to scratch with your fingernail?
2. Which rocks scratched the glass?
3. Which rock is the hardest?

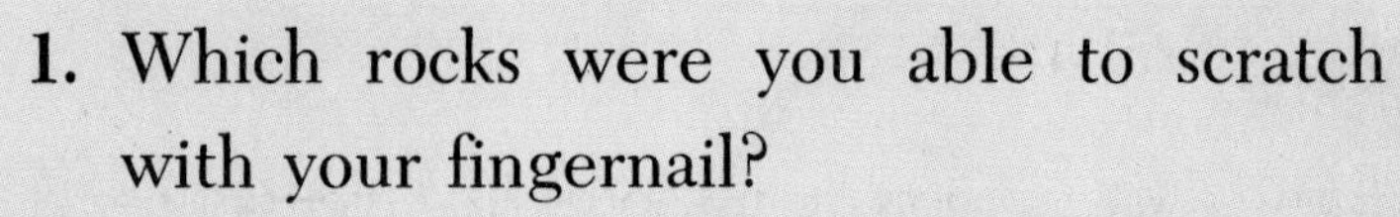

Some rocks break apart easily. They break into flat sheets. Pieces of flat rocks are used as steps in a yard.

Some rocks can be polished to look smooth and shiny. These rocks are used to make the outsides of buildings.

Section Review

Main Ideas: Rocks are used for many things. You decide how a rock can be used by its properties. Hard rocks are not used in the same ways that soft rocks are.

Questions: Answer in complete sentences.

1. What are some ways rocks are used?
2. Pretend that you have found a pretty rock. When you pick up the rock, your fingernail makes a small scratch on it. Is it a hard rock or a soft rock? How can you tell?
3. Give an example of a hard rock and a soft rock.
4. How could each rock you named be used?

CHAPTER REVIEW

Science Words: The clues in column B will help you unscramble the words in column A. Write your answers on a separate sheet of paper.

Column A	Column B
1. NOAMDID	The hardest mineral
2. TARZUQ	A common mineral that is found in many colors
3. TEOLESMIN	A soft rock
4. DERSHANS	A property of rocks and minerals found by using the scratch test
5. INLAMRES	The materials rocks are made of
6. RPOYTERP	Something you can observe, such as color, shape, or feel
7. TEHILA	A mineral that tastes salty

Questions: Answer in complete sentences.

1. What are three properties of rocks?
2. What property of rock makes it useful for steps in a yard?
3. How is a mineral different from a rock?
4. If someone gave you a rock, how could you find its hardness?
5. A rock scratches glass. What does this tell you about the rock?

CHAPTER 5

CHANGING ROCKS

5-1. Weathering

Suppose a bird flew over this rock a long time ago. It dropped a seed on the rock. A tree now grows in a crack of the rock. How could the tree and its roots change this rock?

When you finish this section, you should be able to:

- ☐ **A.** Explain how rocks are changed by *weathering*.
- ☐ **B.** Give some examples of *weathering*.

Weathering: Breaking of rocks into smaller pieces by rain, ice, and plants.

Rocks do not always stay the same. Look closely at this mountain. On its sides and at the bottom are smaller rocks. They have broken off the mountain. The rocks of the mountain are being changed. They are being broken down. Breaking rocks into smaller pieces is called **weathering**. How could *weathering* happen here? Let's find out.

These are tiny plants. They grow on rocks. Many of these plants are growing on the mountain rocks. The rocks are eaten away by the plants. This causes the rocks to break apart. The next time you are near rocks, look for these plants.

Many rocks have cracks in them. When it rains, water can flow into the cracks. During the winter, the rocks get very cold. The water in the cracks freezes. The frozen water makes the cracks bigger. Do you know why?

Look at the picture below at the left. The glass bottle is filled with water. The picture at the right shows the bottle after it was put in a freezer. The glass has cracked. When the water froze, it pushed out. The same thing happens in rocks. Water in rocks pushes out when it freezes. This causes the rocks to break.

ACTIVITY

How are rocks broken down?

A. Gather these materials: drinking straw, 3 jars, water, vinegar, and rocks.

B. Fill 2 jars half full of water. Label 1 jar WATER. Fill the third jar half full of vinegar. Label this jar VINEGAR.

C. Blow through the straw into the jar of water without a label. The air you blow out contains carbon dioxide. Now label this jar CARBON DIOXIDE.

D. Put a few rocks in each jar. Keep 1 rock aside.

E. Put the jars aside. Look at them tomorrow.

F. Take the rocks out of the jars.

1. What changes can you see in the rocks?
2. How did the vinegar affect the rocks?
3. How did the carbon dioxide and water affect the rocks?

CARBON DIOXIDE
WATER
VINEGAR

Look at the picture at the top of page 73. This hole in the ground was caused by weathering. It was made by rainwater. Rainwater picks up a gas called **carbon dioxide** (**kahr**-bun dy-**ahk**-side). When the rainwater and *carbon dioxide* mix, they can change rocks. The water goes into the

Carbon dioxide: A gas in the air.

cracks when it falls to the ground. It goes under the ground. The rocks under the road are weathered. The rainwater makes the rocks break apart.

Section Review

Main Ideas: Rocks are broken down by weathering. Small plants and roots of plants can break rocks apart. Ice and rainwater can weather rocks.

Questions: Answer in complete sentences.

1. What is weathering?
2. Give an example of weathering. How does it change rocks?
3. How can a tree change a rock?
4. Look at this crack in a sidewalk. How could the sidewalk be weathered?

5-2. Soil

This person is using a machine to turn the soil. This is done before seeds are planted. Where did the soil come from? What is it made of?

When you finish this section, you should be able to:

- A. Explain how *soil* is formed.
- B. Name the three layers of *soil*.
- C. Describe the properties of *soil*.

Soil: Tiny pieces of rocks and minerals.

Soil is made from pieces of broken rocks and minerals. It takes a long time to make *soil*. In 500 years nature makes about 2 1/2 centimeters (1 inch) of soil.

A side view of soil is shown below. You might find this soil at the side of a road. Look from the bottom up. At the bottom are rocks. The rocks get smaller as you go up. At the top the soil is dark. There are roots in the soil.

Soil forms in three layers. The top layer of soil is called **topsoil**. It is usually dark. It has twigs and leaves in it. This is where plants grow.

Topsoil: The top layer of soil.

Below the top layer is the **subsoil**. It is usually lighter than the *topsoil*. It can be light brown, yellow, or red. It contains pebbles, sand, and clay.

Subsoil: The layer of soil beneath the topsoil.

The bottom layer of soil is called **bedrock**. It contains large pieces of rock.

Bedrock: The bottom layer of soil; composed of rocks.

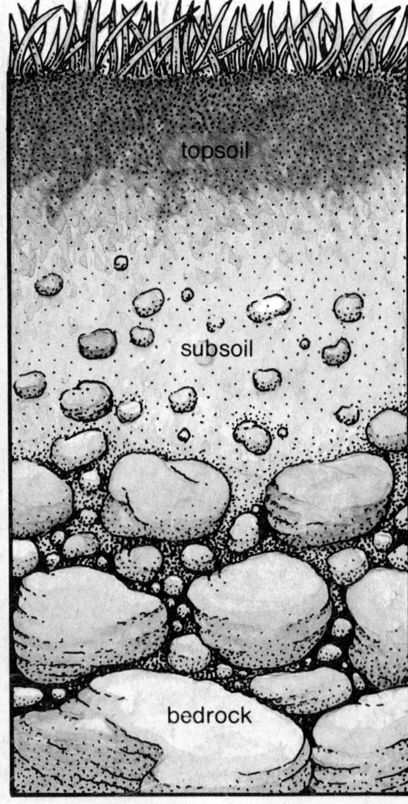

Sandy soil: Does not hold water well.

Clay soil: Holds water very well.

Loam: Soil with sand, clay, and humus.

Humus: Rotting animals and plants in soil.

Soil is important to plants, animals, and people. Plants grow in soil. People eat food that is grown in soil. Some animals live in soil.

Soils look different because they are made of different materials. Some soils are mostly sand. These **sandy soils** do not hold water very well. Most sand is small pieces of quartz.

Another type of soil is **clay soil**. *Clay soil* is made up of very small pieces of a mineral other than quartz. Clay soil holds water very well. When you put water on clay soil, the soil feels very smooth.

Another type of soil is **loam** (**lowm**). *Loam* has sand, clay, and **humus** (**hyoo**-mus) in it. *Humus* is made up of parts of dead plants and animals. These make the soil dark.

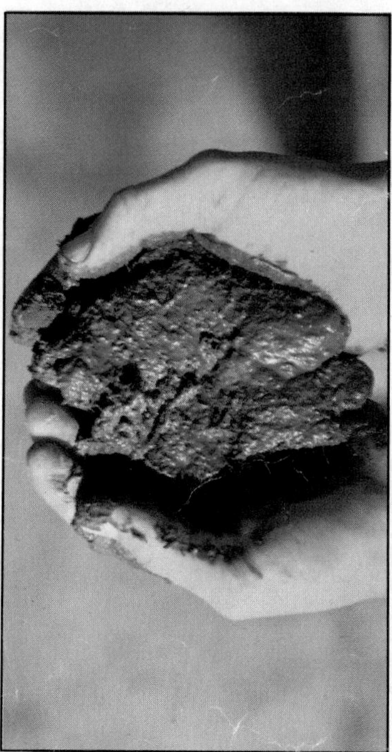

ACTIVITY

What materials make up soil?

A. Gather these materials: 1 bag of topsoil, 1 bag of subsoil, 1 bag of bedrock, hand lens, and newspaper.

B. Pour the bedrock on the newspaper. Describe this part of the soil.

C. Repeat step B for the subsoil and the topsoil.

1. How are the 3 parts of the soil different?
2. How is the subsoil made?
3. How is the topsoil made?

Section Review

Main Ideas: Soil is made from pieces of broken rock. Parts of dead plants and animals are found in some soils. The three layers of soil are topsoil, subsoil, and bedrock.

Questions: Answer in complete sentences.

1. How is soil formed?
2. How are the three layers of soil different from each other?
3. What is loam?
4. You want to drain water from a ball field. Would you put sand or clay under the field?

5-3. Erosion

This would not be a very good place to be during a flood. Too much rain fell here. The water of the river has spilled onto the land. Everything in the path of the water could have been carried away. Do you know what causes floods? Can they be stopped?

When you finish this section, you should be able to:

- **A.** Explain how moving water and wind erode the land.
- **B.** Describe how *erosion* can be stopped.
- **C.** Explain what *conservation* is.

There are rivers all over the United States. They carry broken rock away. The pieces of rock they carry away look like this. This material is called **sediment** (**sed**-uh-ment). Tons of *sediment* are carried away by rivers. The carrying away of sediment is called **erosion** (ih-**roh**-zhun).

A river looks like the drawing shown below at the left. If the water stays inside the banks of the river, everything is fine. But at times water spills over the banks. This is called a flood.

To stop floods, walls are built on the sides of the river. This keeps the water inside the banks. Floods can do a lot of harm. For example, crops may be pulled out of the ground. Homes may be swept away. Floods also wear away the topsoil. But floods can also be good for farmland. The flood waters add minerals to the soil. This is good for the plants growing in the soil.

Sediment: Broken rock that is carried away by water and wind.

Erosion: Carrying away of rocks and soil.

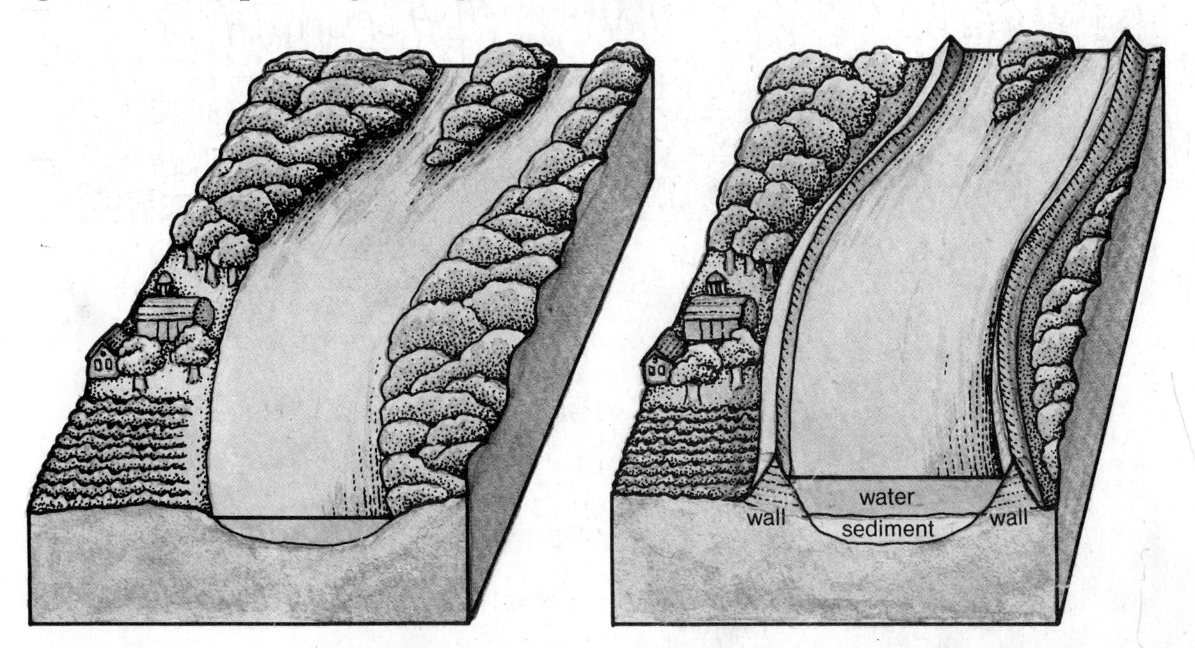

Do you like to watch waves when you are at the beach? Waves of water sometimes crash into the rocks. This wears down or erodes the rocks. The picture at the left shows how a beach can be eroded during a storm.

The wind can erode soil. In some places, the wind blows much of the time. If the soil is dry, the wind can pick it up and carry it away. If grass or other plants are not growing in the soil, the soil can be eroded.

There will always be erosion. Do you know how erosion can cause harm? Soil can be carried away. Plants and trees can be pulled out of the soil. People should try to take care of the land and plants. Taking care of them is called **conservation** (**kon**-sur-**vay**-shun). Planting trees on a hillside is a type of *conservation.* What types of conservation are shown in these pictures?

Conservation: Ways of taking care of the land.

ACTIVITY

How can erosion be stopped?

A. Gather these materials: 2 trays, sandy soil, water, and sprinkler.

B. Fill both trays half full of soil. Make the soil in each tray into a hill.

C. On 1 hill make a series of flat areas that go around the hill. Look at the picture.

1. Which hill do you think will erode faster?

D. Using the sprinkler, slowly pour water over one hill. Then pour water over the other hill.

2. What did you observe?

3. What is one way to slow erosion?

Section Review

Main Ideas: Moving water and wind can erode the land. They carry away broken rock, called sediment. Conservation helps slow down the erosion of the land.

Questions: Answer in complete sentences.

1. How does water erode the land?
2. What is sediment?
3. Does flooding always cause harm?
4. How can wind change the land?
5. Name one type of conservation.
6. How can erosion cause harm?

CHAPTER REVIEW

Science Words: List the letters **a** through **k** on paper. Write the correct word from the list next to each letter.

erosion	**bedrock**	**sandy**
clay	**soil**	**weathering**
subsoil	**conservation**	**humus**
sediment	**topsoil**	

___a___ is made from pieces of broken rocks. The way these rocks are broken up is called ___b___. The three layers of soil are ___c___, ___d___, and ___e___. ___f___ soil does not hold water very well. But ___g___ soil is made up of very small pieces of minerals. It holds water well. There is clay, sand, and ___h___ in loam.

Tons of ___i___ are carried away by rivers. This is a type of ___j___. ___k___ means taking care of the land.

Questions: Answer in complete sentences.

1. Give three examples of how rocks can be weathered.
2. What is an example of something being weathered near your home?
3. What are the three layers of soil?
4. What is one property of each layer of soil?
5. What is the difference between sandy soil and clay soil?
6. How can moving water affect soil?
7. What is a way to stop wind erosion?

CHAPTER 6

FOSSILS

This person is carefully cutting into rocks. He is looking for fossils. He does not want to break the fossils. Do you know what fossils are? Have you ever looked for fossils?

6-1. Fossil Hunting

When you finish this section, you should be able to:

- ☐ **A.** Explain what a *fossil* is.
- ☐ **B.** Describe ways that *fossils* are made.

In this rock, there are many things that look like shells. These shells are examples of **fossils** (**fahs**-silz). *Fossils* are things that are left from plants and animals that lived a long time ago.

Let's find out how these shells became fossils. A long time ago, the animals that lived in these shells were alive. They lived at the bottom of the ocean. The animals in the shells died. The shells became covered with sediment. As the years passed, the sediment hardened into rock. Even when the shells were no longer there, copies of them were still in the rock.

Fossils: Things left from the leaves, bones, or shells of plants or animals that lived long ago.

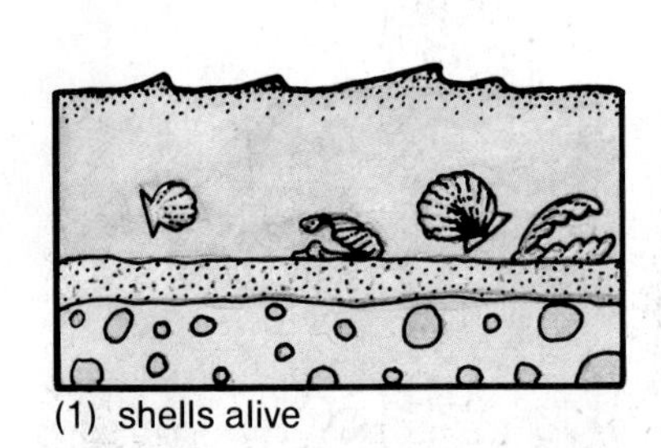

(1) shells alive

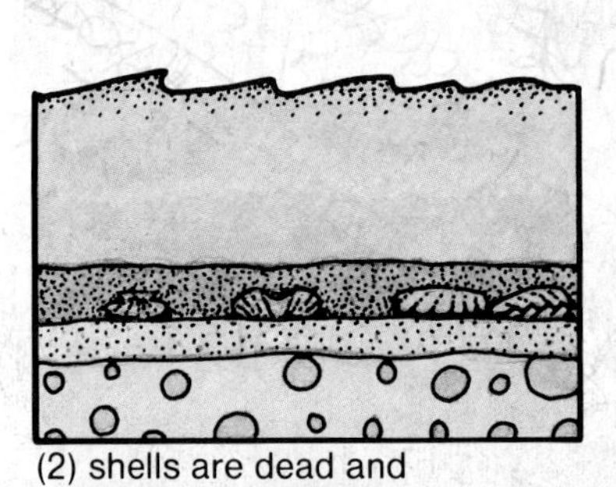

(2) shells are dead and covered by sediments

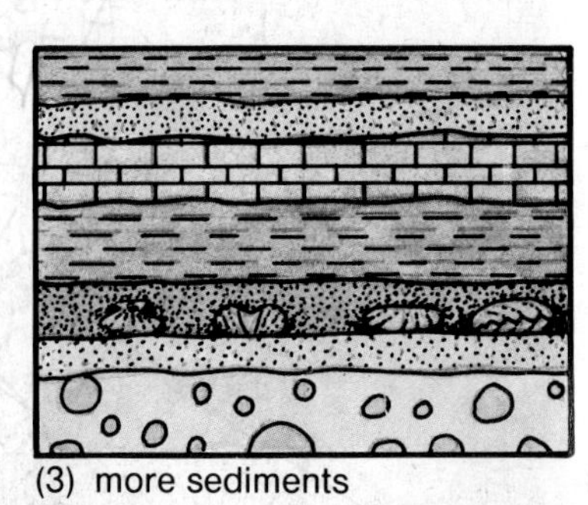

(3) more sediments

(4) sediments lifted

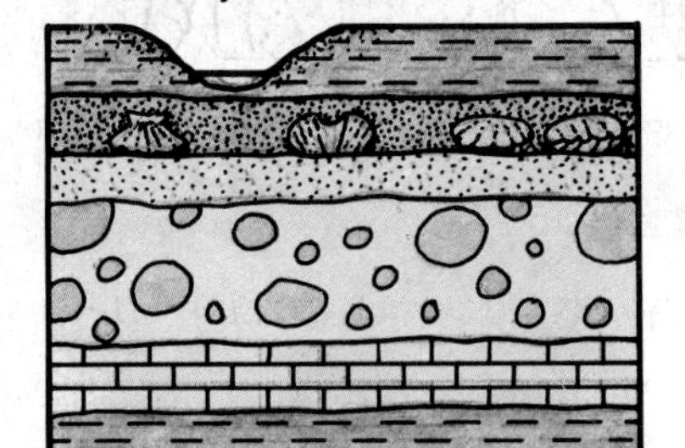

(5) erosion of sediments

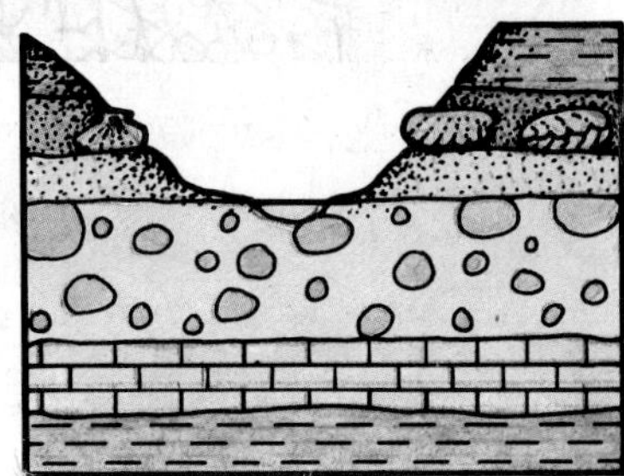

(6) copies of shells can be seen

There are other types of fossils. Fossils can be the teeth or bones of animals. They can also be prints such as the footprints of animals. Fossil prints of plants have also been found.

Some of the best fossils have been found in tar pits. Animals were caught in the sticky tar. They could not get out so they died there.

Other types of fossils have been found in ice. Giant animals that lived near the ice were trapped. Their bodies were frozen. They were found many years later.

Fossils have also been found in amber (**am-ber**). Amber is the sticky material from pine trees. Insects can be caught in the sticky material. After many years, the sticky material hardens into amber.

Activity

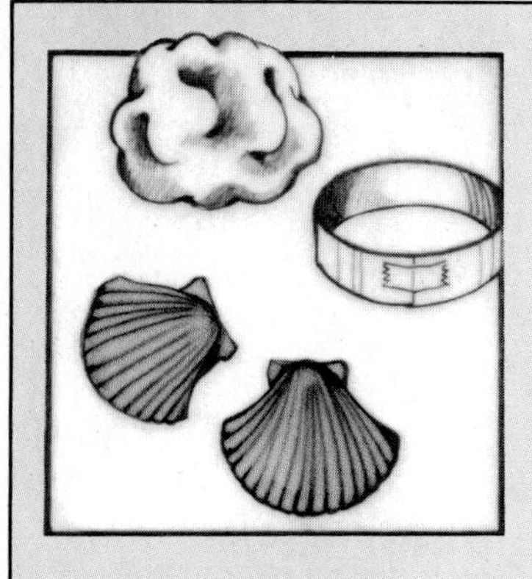

How are fossils made?

A. Gather these materials: several small shells, clay, plaster mix, bowl, water, petroleum jelly, narrow card, and empty soup can.

B. Make a flat circle from a piece of clay. The circle should be a little bigger than the bottom of the soup can.

C. Press the shells into the clay. Then gently remove the shells. Rub some petroleum jelly over the clay.

D. Make a ring with the card that will fit around the print of the shells. Push the card into the clay.

E. Make a thin mixture of plaster and water. Then pour it on the clay. It should be about 2 centimeters deep.

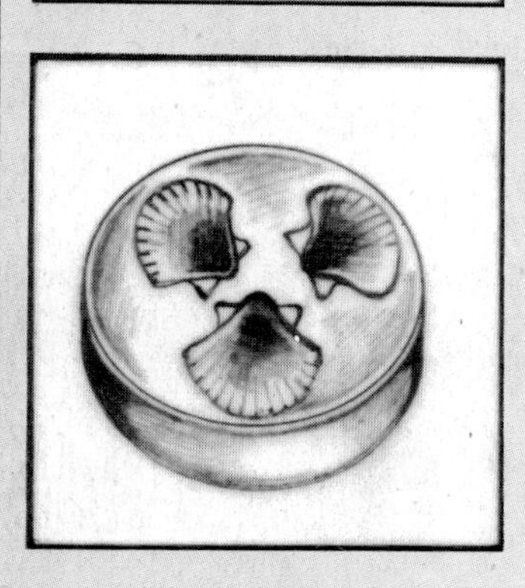

F. Let the plaster harden for about 30 minutes.

G. Carefully peel the clay from the plaster.

H. Compare the "fossils" to the shells.

1. How are the "fossils" like the shells?
2. How is making shell "fossils" similar to the way real fossils are made?

Section Review

Main Ideas: Fossils are things that are left from plants and animals that lived long ago. Fossils can be found in rocks. Fossils have also been found in tar pits and ice.

Questions: Answer in complete sentences.

1. What is a fossil?
2. What are three ways fossils can be made?
3. Where have some of the best fossils been found?

James A. Jensen

James Jensen is known to many people as "Dinosaur Jim." He is a famous fossil hunter. He works as a scientist at Brigham Young University in Utah. But he spends most of his time hunting for fossils.

James Jensen talks to farmers and rock collectors. These people help him look for the best fossil spots. He does not dig unless he sees a bone that could be a fossil. Mr. Jensen always digs carefully. He takes the fossil back to his lab. Then, he studies it more carefully. When he uncovers a fossil he thinks, "Imagine, this is the first sunlight to shine on this animal since it was buried. And I am the first human to see it."

6-2.

Fossil Stories

This is a fossil of an animal. Animals like this lived a long time ago. They probably crawled around on the bottom of the ocean. They can no longer be found as living things. They are only found in rocks as fossils. What do you think they were like? What do you think they ate?

When you finish this section, you should be able to:

- **A.** Explain how fossils help us learn more about the past.
- **B.** Describe what life might have been like a long time ago.

Fossils can tell us what life might have been like a long time ago. We are going to take a trip back in time. Let's begin.

A long time ago life was very simple. Most animals and plants were single cells. They might have looked like the picture below. Rocks have been found with fossils of single-celled organisms.

As time went on, life on earth changed. A very long time ago most things lived in the sea. Fossils of this ancient life have been found. These fossils help scientists learn more about what ancient life might have been like.

There were thick forests during this time. When the plants in the forests died, they piled up. As time went by, layers of rock covered the dead plants. The heavy rocks pressed the plants together. This changed the plants into **coal**. The *coal* we use today was made a long time ago.

Coal: Rock-like material made when plants are pressed together.

Fossils tell more stories. The picture above on the right shows a large **reptile** that lived on earth a long time ago. *Reptiles* are cold-blooded animals such as turtles and snakes. But a long time ago the reptiles were different. Look at the picture again. How is this reptile different from a turtle?

Reptile: A cold-blooded animal such as a turtle or snake.

ACTIVITY

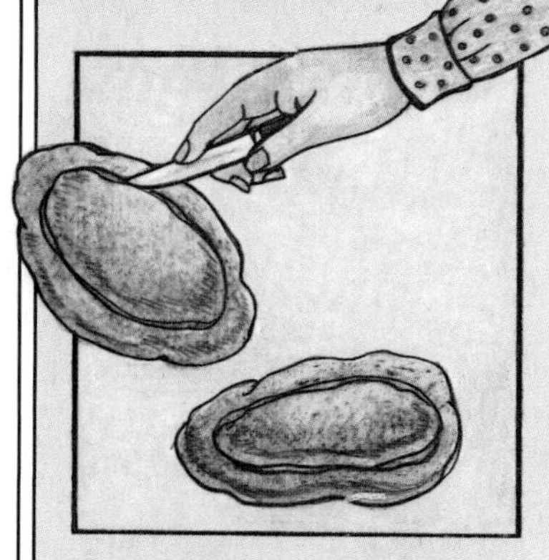

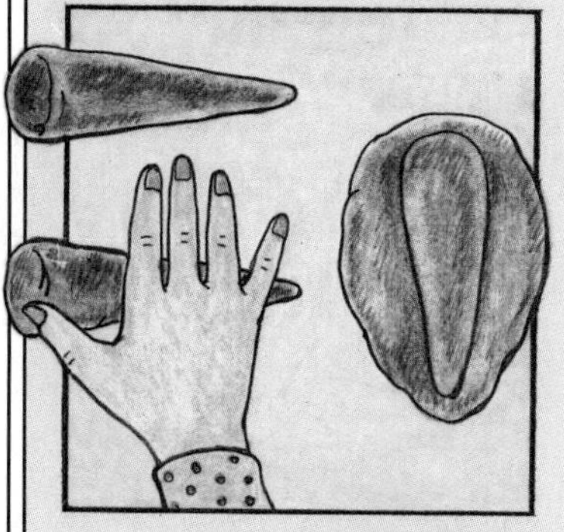

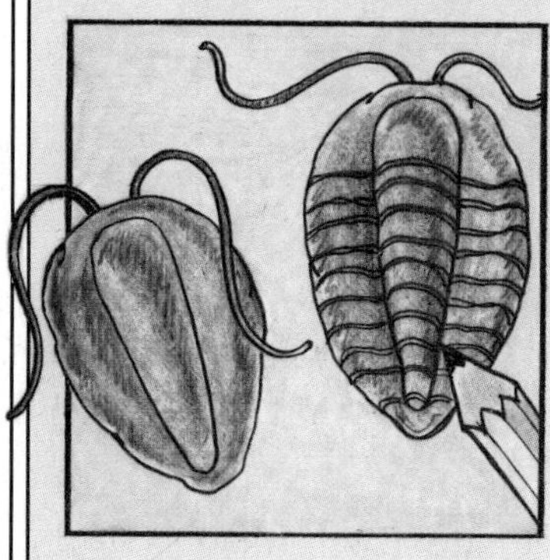

What might life have been like a long time ago?

A. Gather these materials: plastic shoe box or glass bowl, sand, clay, carrot tops, water plants, plastic knife, and pipe cleaner.

B. Roll the clay out flat. Then, using a plastic knife, cut out an egg shape. This is the body of a trilobite (**try**-luh-**bite**), an animal that lived long ago. The picture in the margin on page 88 shows a fossil of a trilobite.

C. Roll another piece of clay into a horn shape that is fatter at one end. Then press this on top of the body of the trilobite.

D. Make scratch marks with the point of a pencil across the body. Use 2 pieces of pipe cleaner for the antennae.

E. Put a layer of sand in the glass bowl. Then plant the carrot tops and the water plants in the sand.

F. Carefully put the trilobite in the sand. Half fill the bowl with water. You have made a model of what life might have been like a long time ago.

1. How could the trilobite become a fossil?
2. Can you describe what life might have been like long ago?

The picture above shows **mammals**. *Mammals* are warm-blooded animals. These animals lived on earth after many of the large reptiles died. Fossils of mammals have been found in rocks in many places. Which of the animals and plants in this picture look a little like those that are alive today?

Mammal: A warm-blooded animal.

Section Review

Main Ideas: Fossils can tell us what life might have been like long ago. Fossils of reptiles and mammals have been found in rocks. Fossils of single-celled animals have also been found.

Questions: Answer in complete sentences.

1. What do fossils tell us about the past?
2. How was coal made?
3. What animals lived on earth after many of the large reptiles died?

6-3.

Dinosaurs

Do you know what dinosaurs are? Do you think a dinosaur could live at your house? Do you think it would be too big? Some dinosaurs were very small. The one shown here was only as big as a dog.

When you finish this section, you should be able to:

- ☐ **A.** Describe what life might have been like for a *dinosaur*.
- ☐ **B.** Name two types of *dinosaurs*.
- ☐ **C.** List reasons why scientists think *dinosaurs* died.

Dinosaur: Means "terrible lizard"; a large reptile that lived long ago.

The word **dinosaur** (**dyn**-oh-sawr) means "terrible lizard." *Dinosaurs* were large reptiles that lived a long time ago.

We know dinosaurs lived because bones of dinosaurs have been found. The picture below shows one of the biggest dinosaurs. This dinosaur is called **Brontosaurus** (**brahn**-tuh-**sawr**-us). The word *brontosaurus* means "thunder lizard." This dinosaur was bigger than seven elephants. Most of the time Brontosaurus lived in rivers and lakes.

Brontosaurus: Means "thunder lizard"; it was a plant-eating dinosaur.

Footprints of dinosaurs have also been found. The picture below on the left shows the footprints of a large dinosaur. These may have been made by Brontosaurus. Perhaps the prints were made in mud. Many dinosaurs stayed in the water.

Fossils of dinosaur eggs have been found. Inside the eggs were the bones of baby dinosaurs.

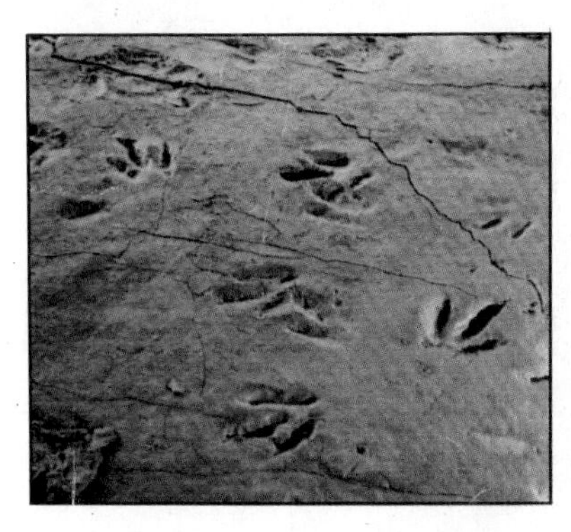

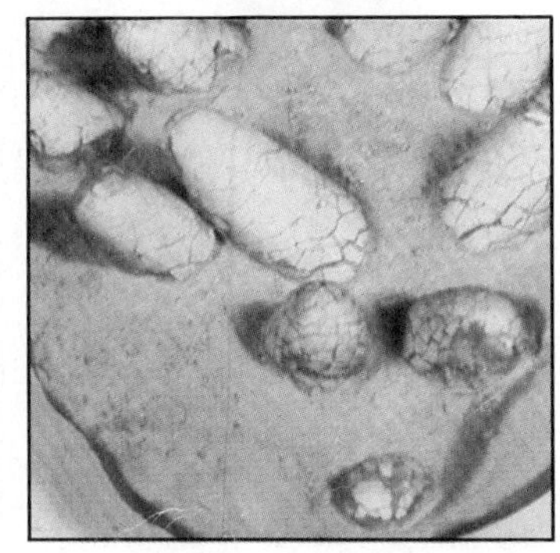

Tyrannosaurus rex: Means "king of the lizards"; it was a meat-eating dinosaur.

Allosaurus: Means "other lizard"; it was a meat-eating dinosaur.

Stegosaurus: A plant-eating dinosaur that had bony plates on its body.

Early dinosaurs were small and moved quickly. Fossil teeth help scientists know that these dinosaurs ate meat. They had plenty of food. Over a long period, they got bigger.

The most powerful of the meat-eating dinosaurs is shown in the margin. It has been named **Tyrannosaurus rex** (tuh-**ran**-uh-**sawr**-us **recks**). *Tyrannosaurus rex* means "king of the lizards." Another fierce meat eater was **Allosaurus** (**al**-uh-**sawr**-us). *Allosaurus* means "other lizard." Allosaurus was feared by many other dinosaurs. Allosaurus is shown at the left below.

Some dinosaurs ate plants. During this time there were many types of plants. Brontosaurus was a plant eater. Another plant eater was **Stegosaurus** (**steg**-uh-**sawr**-us). *Stegosaurus* means "cover lizard." Look at its body. It is covered with bony plates. Some scientists think these plates protected it from meat-eating dinosaurs. As you can see, plant-eating dinosaurs were big.

Dinosaurs no longer live on the earth. They died a long, long time ago. Scientists do not know exactly why.

One reason might have been that the type of food they ate changed. Some think the plants changed. The plant-eating dinosaurs could not eat these new plants. Without food the plant eaters could not live. Do you know why this would also affect the meat-eating dinosaurs? The meat-eating dinosaurs needed the plant-eating dinosaurs for food.

Some scientists think the weather changed. Dinosaurs liked warm weather. If the weather became colder, the cold might have caused the dinosaurs to die. Another reason why dinosaurs died might have been disease.

As you can see, there could have been many reasons why the dinosaurs died. But what if the dinosaurs had lived on? What would it be like to have dinosaurs alive today?

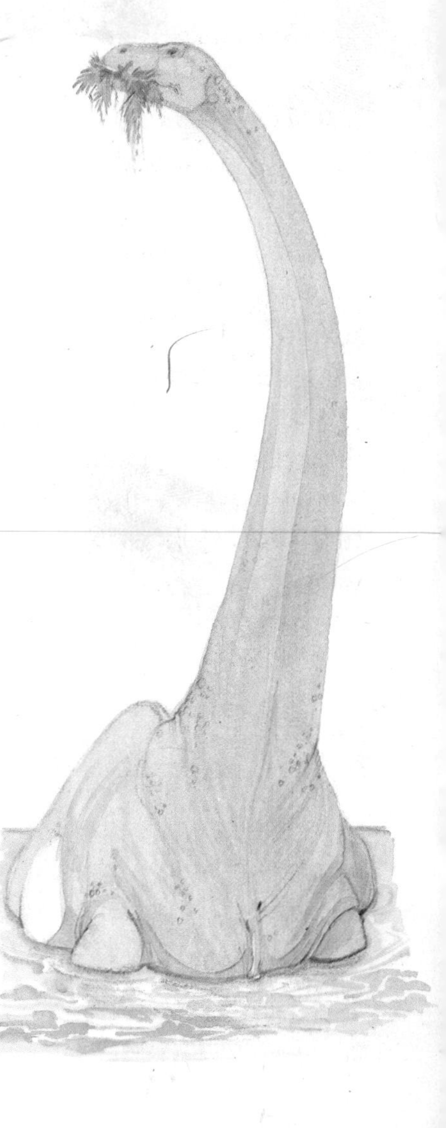

ACTIVITY

What were some of the dinosaurs?

A. Gather these materials: books on dinosaurs.

B. Make a copy of this chart.

DINOSAUR CHART		
Name of Dinosaur	Plant or Meat Eater?	Other Facts

C. Look at the dinosaur shadows. Find out more about these dinosaurs by reading the books and your text. Then complete the chart for each dinosaur.

Section Review

Main Ideas: Dinosaurs were reptiles that lived a long time ago. There were plant-eating and meat-eating dinosaurs. All of the dinosaurs died a long time ago.

Questions: Answer in complete sentences.

1. What are the two types of dinosaurs? Give an example of each.
2. How do we know that dinosaurs once lived on the earth?
3. Were early dinosaurs large or small?

CHAPTER REVIEW

Science Words: Write the sentences below on paper. Fill in the blanks with the correct words from the list.

dinosaur	**ancient life**	**Stegosaurus**
Tyrannosaurus rex	**reptiles**	**mammals**
fossils	**coal**	

1. ______ can be the teeth, bones, or footprints of animals.
2. A ______ is a large reptile that lived long ago.
3. Turtles and snakes are ______.
4. ______ is made when dead plants are pressed together.
5. The king of the dinosaurs was ______.
6. ______ are warm-blooded animals.
7. ______ was a plant-eating dinosaur with bony plates.
8. Fossils of one-celled organisms show us what ______ may have been like.

Questions: Answer in complete sentences.

1. What does a fossil tell us?
2. Describe one way that a fossil is made.
3. What is amber?
4. What is the difference between a reptile and a mammal?
5. What does the word Brontosaurus mean?

INVESTIGATING

Discovering differences among rocks and minerals

A. Gather these materials: 1 card, small pieces of rocks and minerals, glue, and a rock and mineral book.

B. You are going to make a rock and mineral card. Match the rocks and minerals with each of the properties listed below. Some of the rocks may have more than one property. For example, a rock that scratches glass may also feel rough.

a. Scratches glass
b. Can be scratched with fingernail
c. Feels smooth
d. Feels rough
e. Has stripes
f. Is shiny
g. Is dull
h. Is a light color
i. Is a dark color
j. Has a very unusual property

C. When you have found a rock or mineral that scratches glass, glue it to the card. Write **a** under it. If it has another property, you should also write the letter of that property under it.

D. Repeat step C for each of the properties **b–j.** Remember to check each rock for more than one property.

E. Look at the book on rocks and minerals. Try to find some of the rocks and minerals on your card. What are their names?

CAREERS

Paleontologist ▶

You have learned how plant and animal fossils help us know about the past. A paleontologist (**pay**-lee-un-**tahl**-uh-jist) is a person who studies fossils. Paleontologists study parts of dead plants and animals. People who are paleontologists have studied biology in college.

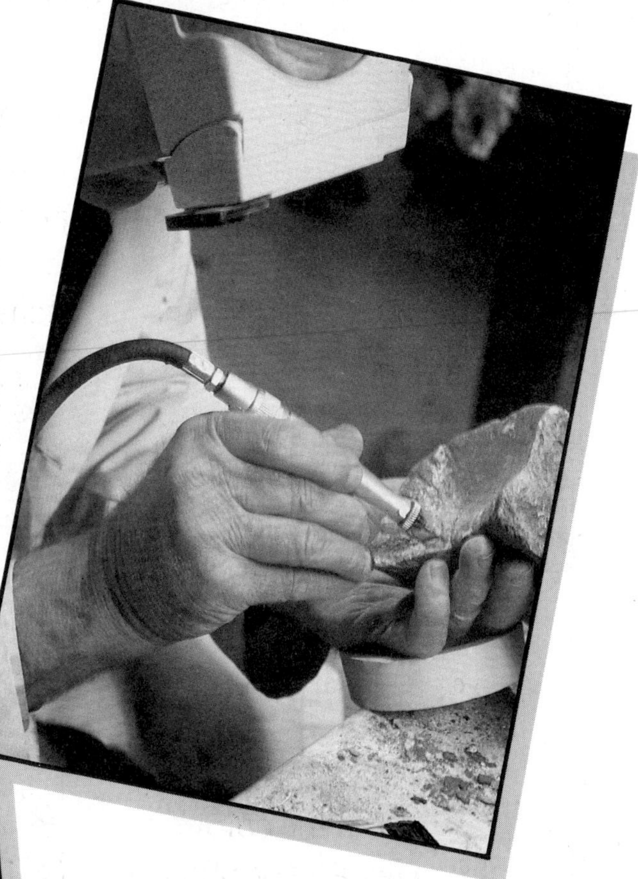

◀ Soil Conservationist

You have learned about rocks and soil. Soil conservationists (kahn-ser-**vay**-shun-ists) may tell farmers and ranchers how to keep water and wind from carrying away their soil. Or they may help farmers decide what crops to grow. A soil conservationist has studied earth science in school.

CHANGES IN MATTER

UNIT 3

CHAPTER 7

MATTER

7-1. Solids, Liquids, and Gases

Pretend you are in an ice-cream store. You ask for an ice-cream soda. You get one that looks like the ice-cream soda on the next page. Look at that soda closely. What is it made of?

When you finish this section, you should be able to:

- **A.** Explain what *matter* is.
- **B.** Name the three forms of *matter*.

All things around you are called **matter**. *Matter* is anything that takes up space and has weight. Look at the picture below. Do you think the ice-cream soda is matter? If you said yes, you are right. Without moving the ice-cream soda, could you put something in its place? If you picked it up, how would it feel? The ice-cream soda takes up space. It has weight.

Matter: Anything that takes up space and has weight.

The three forms of matter are solids, liquids, and gases. We can find each form in the strawberry ice-cream soda. Of course, every ice-cream soda has ice cream in it. The ice cream is a solid. The strawberries are solids, too. The soda is a liquid (**lih**-kwid). Do you know what the little bubbles in an ice-cream soda are? The bubbles are a gas. The air in your classroom contains many gases. Usually, you cannot see gases.

The three forms of matter are shown in the above picture. Can you name the solids, liquids, and gases? The wood, pencil, book, and eraser are solids. The milk, juice, and water are liquids. The air in the balloon and in the jar contains many gases.

If you put a block of wood in your desk, would the shape of the wood change? What would happen if you put the wood in your bathtub? Would its shape change? A solid keeps its shape no matter where it is placed.

Look at the picture below. What shape is the colored water in the tall glass? What shape is the water in the flat dish and in the fish bowl? Liquids change shape. The shape of a liquid depends on the shape of its container.

There is air in the bottle shown here. Air contains many gases. Do the gases fill the bottle? A gas takes on the shape and size of its container.

Look at the sign below. It is filled with a gas called neon. When electric current passes through neon, it gives off a bright red light. How is the shape of the gases in the bottle different from the shape of the gas in the sign?

Section Review

Main Ideas: Matter takes up space and has weight. The three forms of matter are solids, liquids, and gases. A solid always keeps its shape. A liquid takes the shape of its container. A gas spreads out and fills its container.

Questions: Answer in complete sentences.

1. What is matter?
2. Which of these are matter: a ball, sunlight, water, air?
3. What are the three forms of matter?
4. What form of matter is each of these items: an ice cube, a crayon, a skateboard, paint?

7-2. Heating Matter

Ed is making a list of things that he can do to keep this snowball from melting. What do you think will happen to the snowball if Ed leaves it on the table for an hour? Have you ever tried to keep a snowball from melting? What did you do?

When you finish this section, you should be able to:

- **A.** Tell how a solid can be changed into a liquid.
- **B.** Tell how a liquid can be changed into a gas.
- **C.** Explain what *melt*, *boil*, and *evaporate* mean.

By the time Ed finished making his list, there was no longer a snowball on the table. There was a lot of water. What happened? The snowball was a solid. It changed into a liquid. When matter changes from a solid into a liquid, we say that it **melts**. Heat causes many kinds of solid matter to *melt*.

Melt: To change from a solid into a liquid.

Look at the picture of the icicle. What form of matter is the icicle? Is it in a solid form, a liquid form, or a gas form? How is the icicle changing? What is causing the icicle to melt? What form of matter is the water?

Imagine placing the water from that melting icicle into a pot. If you were to heat the water, a change would take place. The liquid would get warmer and warmer. Soon, the liquid would get so hot it would start to bubble.

In the above picture, do you see a cloud? What is that cloud? It is **steam** (**steem**). *Steam* is tiny drops of liquid water in the air. Before long, the liquid will be gone. The steam will also be gone. Do you know why? First, tiny drops of liquid go into the air as steam. The steam disappears as the drops of liquid turn into gas. The gas cannot be seen. The changing of matter from a liquid into a gas is called **evaporation** (ih-va-poh-**ray**-shun). Heat causes most liquids to *evaporate*. If the change is happening quickly, the liquid is **boiling**. A *boiling* liquid bubbles. A liquid does not have to boil to evaporate.

Steam: Tiny drops of water in the air.

Evaporation: The changing of a liquid into a gas.

Boil: To change quickly from a liquid into a gas.

Look at the picture below. Matter is changing form here. Can you explain what is happening? What do you see over the ice? A hot rod is melting the ice.

ACTIVITY

Can you make an ice cube melt quickly?

A. Gather these materials: several ice cubes and a plastic bag.

B. Think of ways to make ice cubes melt. Then, tell your ideas to the teacher.

1. Did your class think of many ideas?

2. Which idea will work best?

C. Your teacher will give you an ice cube in a plastic bag. Use one of your ideas about how to melt ice. When your ice melts, tell your teacher.

3. Did your ice cube melt the fastest?

4. Why did some ice cubes melt faster?

Have you ever hung wet clothes outside on a warm day? What happened to the water in the clothes? The warm air caused the water to evaporate. The water became a gas. So, the clothes dried.

When you go swimming, you get wet. When you get out of the water, you sometimes sit in the sun. If you do, you become dry. Do you know where the water goes?

Section Review

Main Ideas: Matter can change form. Heat causes many types of solid matter to melt. Heat causes many types of liquids to evaporate.

Questions: Answer in complete sentences.

1. What happens to solid matter when it melts?
2. What is evaporation?
3. What happens to liquid matter when it evaporates?
4. After a rainstorm, there is often a big puddle on the sidewalk. Later, that puddle is gone. What happened to it?

7-3. Cooling Matter

Did you ever write on a bathroom mirror with your finger? The next time you step out of a warm shower or bath, look at the mirror. Probably, you will not be able to see yourself. The mirror will look foggy. Try to write your name on the mirror.

When you finish this section, you should be able to:

- **A.** Tell how a gas can be changed into a liquid.
- **B.** Tell how a liquid can be changed into a solid.
- **C.** Explain what *condensation* and *freezing* mean.

In the last section, you learned what happens to matter when it is heated. Now, you will learn what happens to matter when it is cooled. Can you guess what that is?

One of the many gases in the air is **water vapor** (**vay**-per). *Water vapor* is always in the air. Water vapor forms when liquid water changes into a gas. If we cooled the air, some of the water vapor would change back into liquid water.

Water vapor: The thing that forms when liquid water changes into a gas.

Why is there water on the outside of the glass shown in the margin? The ice cubes made the glass cold. The cold glass cooled the water vapor in the air near the glass. Cooling the water vapor made it change from a gas back into a liquid. Changing matter from a gas into a liquid is called **condensation** (kahn-den-**say**-shun). Cooling causes most gases to *condense*. The cold glass cooled the gas in the air. The gas condensed.

Condensation: The changing of a gas into a liquid.

ACTIVITY

What happens on the outside of a metal can when there is ice and water in it?

A. Gather these materials: shiny metal can, water, and ice cubes.

B. Place the ice cubes and water in the can. Then, watch the outside of the can for a few minutes.

1. What do you see on the outside of the can?
2. Explain why you think this happened.

The fog you see on a bathroom mirror is made of tiny drops of water. The mirror is cold. The water vapor in the air is cooled when it hits the mirror. The gas becomes a liquid. The liquid makes a thin covering on the mirror.

You breathe out many gases. One is water vapor. You cannot see that gas. But you can see it condense if you are outside on a cold day. Look at the picture above. What is the dog breathing out? The dog is breathing out water vapor. When the water vapor meets the cold air, it changes into tiny liquid drops. The tiny drops form the cloud in front of the dog's mouth.

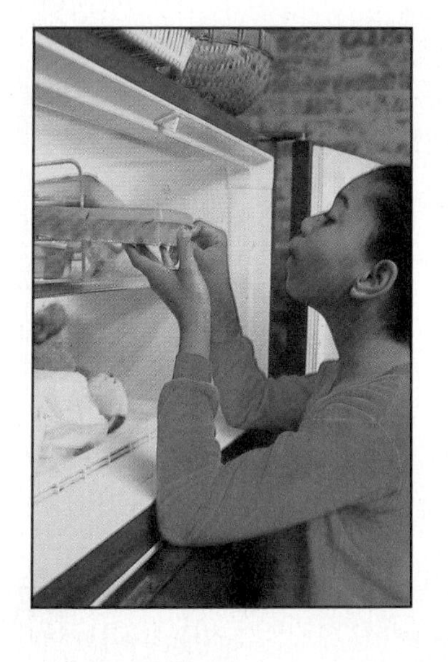

Freeze: To change from a liquid into a solid.

A liquid can also change into a solid. Probably, you have seen this happen many times. Can you think of an example? How can you change liquid water into a solid? Put the water into a freezer. The liquid water will become solid. It will turn into ice. When water changes from a liquid into a solid, we say that it **freezes**. *Freezing* means to change from a liquid into a solid. Cooling causes liquids to become solids.

Section Review

Main Ideas: Matter can change form. When a gas changes into a liquid, it condenses. When a liquid changes into a solid, it freezes.

Questions: Answer in complete sentences.

1. What happens to liquid matter when it freezes?
2. What happens to gas when it condenses?
3. It is very cold outside. It is warm inside. Why is there ice on the inside of the window? Hint: The glass is very cold.

Eugenie Mielczarek

Dr. Eugenie Mielczarek (mee-el-**char**-ek) teaches physics at George Mason University. Dr. Mielczarek also does research work. Like many other physicists, she has an interest in biophysics. Biophysics is a combination of physics and biology.

When she was in high school, one of Dr. Mielczarek's teachers urged her to go into physics. After reading a physics textbook, she knew she wanted to enter the field. After she left college, she worked at the National Bureau of Standards in Washington, D.C. Dr. Mielczarek was awarded a Ph.D. in physics in 1961. After that she was an assistant research professor at Catholic University.

7-4.

Matter Changes Size

The girl in the picture cannot open the jar. She is trying with all her might. Has this ever happened to you? There is an easy way to open a stuck lid. You will learn what that way is.

When you finish this section, you should be able to:

- **A.** Tell what happens to the size of something when it is heated.
- **B.** Tell what happens to the size of something when it is cooled.

Expand: To get larger.

Many things get larger when heated. When matter gets larger, we say that it **expands** (eks-**pandz**). Most gases, solids, and liquids *expand* when heated.

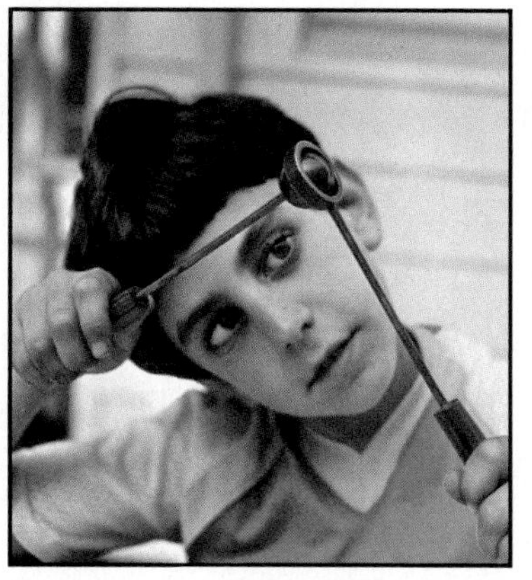

Look at the pictures above. In the first picture, the boy can easily fit the ball through the ring. But, after he heats the ball, it will not fit. Do you know why? The ball expanded when heated. It became too large to fit through the ring. In the last picture, the boy is placing the ball in a bowl of ice-cold water. The ball will easily fit through the ring again. Do you know why? The cold water cooled the ball. Cooling caused the ball to get smaller. Many things get smaller when cooled. When matter gets smaller, we say that it **contracts** (kon-**trakts**). Most gases, solids, and liquids *contract* when cooled. Water is different from most matter, however. When cooled to a solid, water expands.

Contract: To get smaller.

Some types of matter expand more than others. Some types of matter contract more than others. If the girl with the jar runs hot water on the jar's lid, it probably will open easily. The hot water will heat the lid. The heat will cause the lid to expand. The lid will expand more than the jar. It will turn easily.

Look at the balloons in the picture. Do you know how they work? The air in the balloons is heated. This makes the air expand. The balloons get bigger. If the weight of a balloon is less than the weight of the air it is pushing aside, the balloon will rise. Balloons can be held down by weights until they are ready to be sent into the air. To return a balloon to the earth, the air is let out of the balloon. As the amount of air in the balloon becomes less, the balloon becomes smaller. Now, the weight of the balloon is more than the air it is pushing aside. So, the balloon comes down.

Have you ever noticed the cracks between the parts of a sidewalk? Do you know why those cracks are there? The cracks allow the sidewalk to expand. Do you know when the sidewalk expands? It expands when it is heated by the sun. This happens most often in the summer. When do you think the sidewalk might contract? It contracts when it cools. This happens mostly at night and in the winter. The cracks get wider. If the sidewalk could not contract and expand, it would break apart.

Section Review

Main Ideas: Matter can expand and contract. When matter expands, it gets larger. Heating matter causes it to expand. When matter contracts, it gets smaller. Cooling causes matter to contract.

Questions: Answer in complete sentences.

1. What happens to matter when it expands?
2. What causes matter to expand?
3. What happens to matter when it contracts?
4. What causes matter to contract?
5. Michael bought a balloon at the circus. He put the balloon in a warm place in his room. An hour later it looked a little bigger. Why did the balloon get bigger? What would have happened if Michael had put his balloon in the freezer?

Chapter Review

Science Words: List the letters **a** through **k** on paper. Write the correct word from the list next to each letter.

water vapor	**expand**	**gases**
liquids	**solids**	**condense**
space	**contract**	**freezes**
melts	**mass**	

Matter is anything that takes up __**a**__ and has __**b**__. The three forms of matter are __**c**__, __**d**__, and __**e**__. When solid matter is heated, it often __**f**__ to become a liquid.

__**g**__ is formed when liquid water changes into a gas. Cooling causes most gases to __**h**__. If a liquid is cooled enough to become a solid, it __**i**__. Heating matter causes it to get larger, or __**j**__. Cooling matter causes it to get smaller, or __**k**__.

Questions: Answer in complete sentences.

1. Why do clothes on an outdoor clothesline dry?
2. Why does ice melt when you put it in a glass of orange juice?
3. What happens to water when you put it in a freezer?
4. What is an easy way to open a jar when its lid is stuck?
5. What are the little bubbles in an ice-cream soda?

CHAPTER 8

HEAT

8-1. Sources of Heat

Have you ever toasted marshmallows over a fire? Did you do this on a cold night? If you did, you know that the heat from the fire kept you warm.

When you finish this section, you should be able to:

- **A.** Name two sources of heat.
- **B.** List four things we burn for heat.

If it were not for the sun, the earth would be a dark, cold place. We get most of our heat from the sun. Even on cloudy days, some heat from the sun reaches us.

Fuel: Anything burned to make heat.

Some places on the earth get more heat from the sun than others. People who live in these places do not have to heat their homes. Do people heat their homes where you live? If so, the heat probably comes from burning **fuel** (**fyoo**-el). Anything that can be burned to make heat is a *fuel.* Fuel is used to heat homes and to cook food. Fuel is used to heat water for washing dishes and bathing. Wood, coal, oil, and gas are fuels. What kind of fuel is used in your home? Is the same kind of fuel used in your school?

Look at the first three pictures above. What type of fuel is making heat?

Most people use wood in their fireplaces. The wood burns. How would you feel if you sat by a fireplace?

To cook on a barbecue, you need charcoal. The charcoal burns. The heat cooks the food. In a home, food may be cooked on a gas stove. Burning gas causes the flame.

The man in the last picture is filling a tank in the house with oil. When the oil is burned, it will heat the air and water in the house.

ACTIVITY

Does heat from the sun affect everything in the same way?

A. Gather these materials: newspaper, black paper, and white paper.

B. Place the folded newspaper on a windowsill in the sunlight. Put the black and the white paper on the newspaper.

C. Wait 15 minutes. Then, touch each piece of paper.

1. Which paper was warmer?
2. What do you think would happen if you used other colors?

Some people use the heat from the sun to heat their homes. The picture below shows a house heated by the sun. Pipes that carry water are under the glass on the roof. The sun heats the water in the pipes. The heated water travels in pipes to the sinks, showers, and bathtubs in the house. The heated water also heats the air in the house.

Heat from the sun can be used to make electricity. Look at the mirrors in the picture above. The mirrors throw the sunlight back onto a boiler at the top of the tower. The heat from the sun turns the water in the boiler into steam. The steam then turns the machinery, which makes electricity.

Section Review

Main Ideas: We get heat from the sun and from fuel. A fuel is anything burned to make heat. Some people use the heat from the sun to heat their homes.

Questions: Answer in complete sentences.

1. What is the source of most of our heat?
2. Some cities burn waste material to make steam. Explain why this makes the waste material a fuel.
3. Name four fuels.

8-2. Measuring Heat

Ann is getting ready for school. She is trying to decide what to wear. How do you think looking out the window will help her decide? Find something in the picture that might be helpful.

When you finish this section, you should be able to:

- A. Read a *thermometer*.
- B. Name two common *temperature* scales.
- C. List the *temperatures* at which water freezes and boils.

Outside Ann's window is a **thermometer** (ther-**mahm**-eh-ter). A *thermometer* measures how hot or cold something is.

A thermometer is shown in the margin. Thermometers measure **temperature** (**tem**-per-uh-cher). *Temperature* is the amount of hotness or coldness in anything. Look closely at the thermometer. What do you see on it? The numbers stand for **degrees** (duh-**greez**). Temperature is measured in units called *degrees*. Lines mark the degrees. Do you see the red liquid in the thermometer? This liquid moves up and down as the temperature changes.

This thermometer has two kinds of scales. The scales are the numbers and lines used for measuring. The one on the left side of the thermometer is the **Fahrenheit** (**fa**-ren-hite) scale. The man who invented it was named *Fahrenheit*. The scale on the right side is the **Celsius** (**sel**-see-us) scale. *Celsius* is the name of the man who invented this scale. Look at the picture. On the Celsius scale, the liquid is at the line marked 28. Therefore, the temperature is 28°C. The small circle means "degree." The **C** at the top of the thermometer stands for Celsius. Look at the picture again. The **F** stands for Fahrenheit. What is the temperature on the Fahrenheit scale?

You learned about freezing and boiling in Chapter 7. The temperature at which matter

Thermometer: Measures how hot or cold something is.

Temperature: How hot or cold something is.

Degrees: Units of temperature.

Fahrenheit and Celsius: Types of thermometer scales.

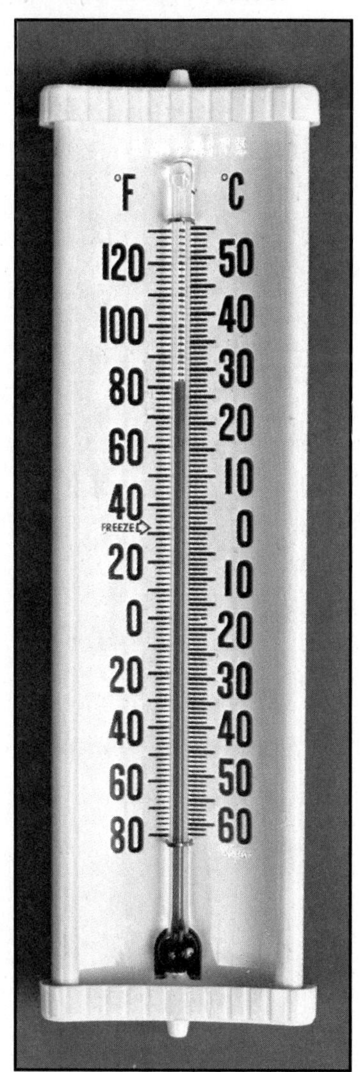

freezes is called its **freezing point**. The *freezing point* of water is 0°C, or 32°F. If the temperature rises above 0°C (32°F) when it is snowing, what will happen? The snow will turn into rain. The temperature at which matter boils is its **boiling point**. The *boiling point* of water is 100°C, or 212°F.

Freezing point: The temperature at which matter freezes.

Boiling point: The temperature at which matter boils.

ACTIVITY

How can I measure temperature?

A. Gather these materials: 1 cup of warm water, ice cubes, paper, pencil, spoon, and Celsius or Fahrenheit thermometer.

B. Place the thermometer in a cup of warm water. When the liquid in the thermometer stops moving, read the temperature. Write it on the paper. Be sure to write °C or °F.

C. Take the thermometer out of the cup. Put ice cubes in the cup. Stir the ice in the water for 1 minute.

D. Put the thermometer in the water again. Write down the temperature of the water.

1. Did you use a Celsius or a Fahrenheit reading?
2. What was the temperature of the warm water?
3. What was the temperature of the cold water?
4. How many degrees did the ice cool the water?

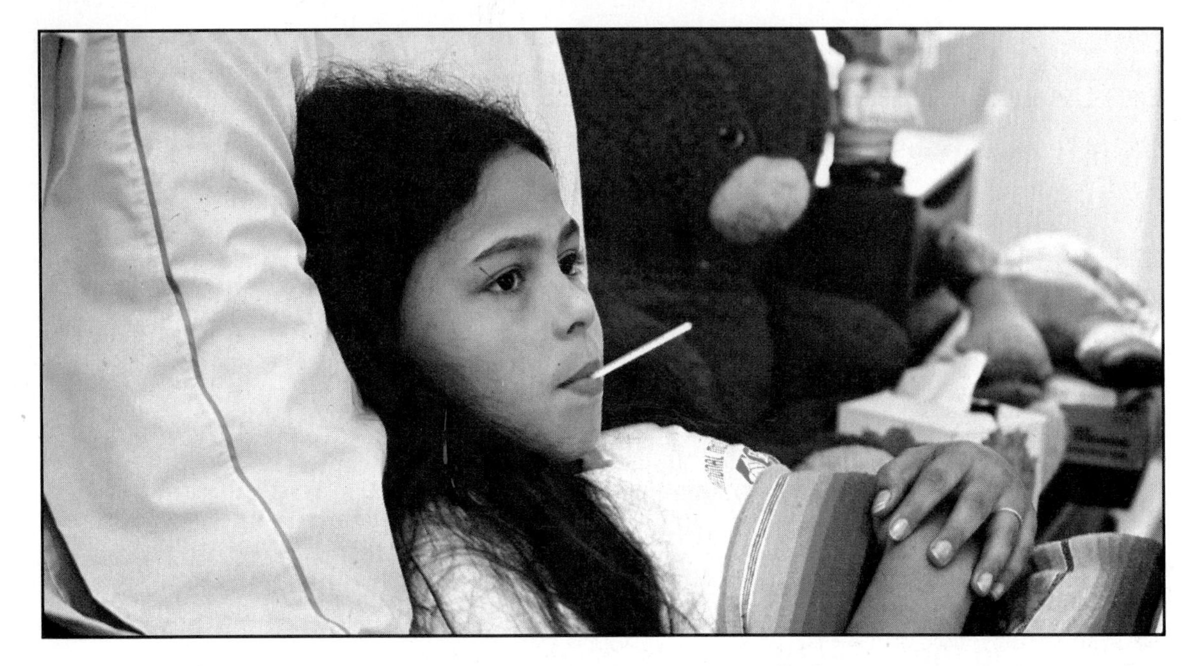

Knowing how to read the temperature can be helpful. The girl in the above picture is sick. She is having her temperature taken. Have you ever had your temperature taken when you were sick? When you are healthy, your temperature is most likely 37°C, or 98.6°F. When you are sick, your temperature goes up.

Do you listen to weather reports on the radio or TV? As shown on page 130, weather reporters often give the temperature readings from both scales. They usually say the Fahrenheit temperature first, and then the Celsius temperature.

The temperature outside affects the way you dress. If the temperature is 2°C (about 36°F), you should wear a coat. If the temperature were 27°C (81°F), you could go swimming. What could you do if it were -6°C (21°F) outside?

Section Review

Main Ideas: Temperature tells how hot or cold something is. Temperature is measured in degrees with a thermometer. Two types of thermometer scales are Celsius and Fahrenheit.

Questions: Answer in complete sentences.

1. What is the temperature on the thermometer shown at the right?
2. Name the two common thermometer scales.
3. What do the numbers on a thermometer stand for?
4. What is the freezing point of water on the Celsius scale?
5. What is the boiling point of water on the Fahrenheit scale?
6. What should your temperature be on the Fahrenheit scale when you are healthy?

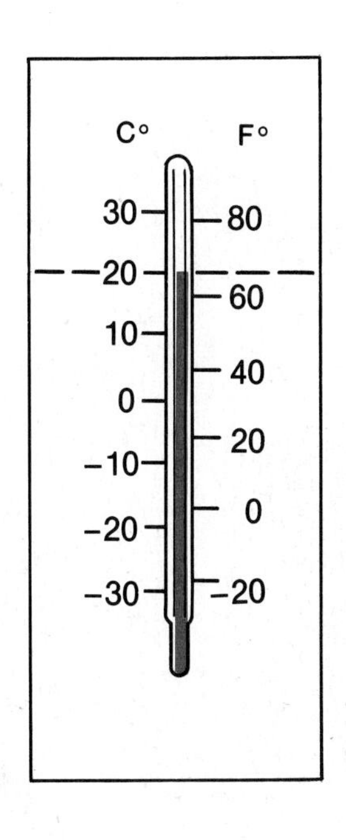

CHAPTER REVIEW

Science Words: The clues in column B will help you unscramble the words in column A. Write your answers on a separate sheet of paper.

Column A	Column B
1. REIFZGNE IPTNO	The temperature at which matter freezes
2. ULEF	Something we burn for heat
3. SLEUCIS	A temperature scale
4. MEHRTEETORM	Used to tell how hot or cold something is
5. REGEEDS	Units used to measure temperature
6. GONLBII TONIP	The temperature at which matter boils
7. NSU	Our main source of heat

Questions: Answer in complete sentences.

1. At what temperature on the Fahrenheit scale will snow turn into rain?
2. What do the numbers on a thermometer stand for?
3. At what temperature on the Celsius scale will water boil?
4. What happens to the liquid in a thermometer outside a window when it gets colder?
5. What kind of clothing would you wear outside if the temperature were 2°C (36°F)?

CHAPTER 9

MATTER IN WATER AND AIR

9-1.

Water Changes Matter

Do you ever swim in the ocean? Do you get sea-water in your mouth? What does it taste like? Do you know why it tastes that way? Did you know that there are many kinds of matter in sea-water?

When you finish this section, you should be able to:

- **A.** Explain how heat helps matter to *dissolve.*
- **B.** Tell what a *solution* is.
- **C.** Explain how *crystals* are formed.

Have you ever made lemonade? How did you make it? First, you squeezed the juice from the lemons. Then, you mixed the juice in water. Did the lemon juice stay on top of the water? Did it sink to the bottom? No, it mixed with the water. You could not tell the lemon juice from the water. The lemon juice **dissolved** (dih-**sahlvd**) in the water. When something mixes with a liquid so that you cannot tell the two things apart, something has *dissolved.*

Dissolve: To mix something with a liquid so you cannot tell the two things apart.

You add sugar to lemonade to make it sweet. Would it be easier to dissolve the sugar in warm water before you put ice in the lemonade? Yes. Heat makes tiny pieces of matter move faster. The tiny pieces of sugar would mix more easily with warm water. They don't mix as easily with cold water.

A **solution** (suh-**loo**-shun) is a mixture formed by dissolving a solid or a liquid in another liquid. Lemonade is a *solution*. Why is it a solution? Ocean water is also a solution. It has a lot of salt dissolved in it.

When a solution evaporates, something interesting happens. The matter that was dissolved forms a solid again. The solid usually becomes a **crystal** (**kris**-tul). Look at the picture below. The *crystals* on the string are sugar crystals.

Solution: A mixture formed by dissolving a solid or liquid in another liquid.

Crystal: A solid that forms when a solution evaporates.

ACTIVITY

Does salt dissolve more quickly in hot water or in cold water?

A. Gather these materials: 2 glass containers, 2 spoons, salt, hot water, and cold water.

B. Fill 1 container with hot water. Then, fill the other one with cold water.

C. Put 1 spoon of salt into each container. Have someone help you, so that you add the salt to both containers at the same time. Stir both solutions at the same speed.

1. In which container did the salt dissolve first?
2. Does heat help salt to dissolve in water?

Section Review

Main Ideas: When matter dissolves in a liquid, it forms a solution. It is easier to dissolve solid matter in warm water than in cold water. When solutions evaporate, matter that was dissolved forms a solid again. That solid is usually a crystal.

Questions: Answer in complete sentences.

1. What is a solution?
2. Name two solutions.
3. Why is it easier to dissolve sugar in warm water than in cold water?
4. What is left when a solution evaporates?

9-2. Air Changes Matter

When you are running, you breathe hard. You need a lot of air. Do you know why? What is in the air that your body needs? Do you know what it does?

When you finish this section, you should be able to:

- **A.** Tell what *oxygen* is.
- **B.** Explain how *oxygen* causes burning and rusting.
- **C.** Explain how *oxygen* is used in your body.

Do you remember what form of matter air is? Air is a mixture of many gases. One of these gases is called **oxygen** (**ahk**-sih-jun). *Oxygen* is the part of the air your body needs. Oxygen is a gas that can change matter.

Oxygen: A gas in the air.

Do you enjoy sitting in front of a fire? Fuel burns when oxygen changes it into something else. Without oxygen, the fire would not burn.

It takes three things to start a fire. There must be fuel, oxygen, and enough heat to start the fuel burning. When fuel burns, it gives off heat and light. The oxygen changes the fuel into other gases. Sometimes there are solids left after fuel burns. Do you know what they are called? They are called ashes.

What does a fire need to burn? If you take any of these three things away, the fire will go out. So, there are three ways to put out a fire. One way is to keep oxygen from getting to the fuel. Campers do this by covering the fire with dirt. Another way to put out a fire is to cool it off with water. The third way is to remove its fuel.

Oxygen also changes other kinds of matter. It causes iron and steel to rust. When oxygen changes metals like copper and iron, it does not make any light. It does make heat. Do you know why you cannot feel the heat? It is because the heat is made slowly.

ACTIVITY

Will a candle burn longer when there is a lot of air or a small amount of air?

A. Gather these materials: pie plate, clock, candle, 500-milliliter jar, 1-liter jar, matches, and water.

B. Stand a candle in the middle of the pie plate. Fill the pie plate with water. Have an adult help you light the candle.

C. Turn the 500-milliliter jar upside down over the candle.

1. How many seconds did the flame burn?

D. Light the candle again. Turn the 1-liter jar upside down over the candle.

2. How many seconds did the flame burn?
3. What was the fuel for the candle's flame?
4. Where was the oxygen that made the fuel burn?
5. Will a candle burn longer if it has more oxygen?

Your body gets oxygen when you breathe. Do you know why you need oxygen? You need oxygen to burn the food you eat. Why do you think you need oxygen to burn food? What do you get from oxygen? When oxygen burns your food, it makes energy. The energy is used to make you grow. You also need energy to work and play.

Section Review

Main Ideas: Oxygen is a gas in the air. It can change matter. It makes energy. Oxygen changes fuel so it can burn. It causes iron and steel to rust. Oxygen changes food in your body so you have energy.

Questions: Answer in complete sentences.

1. What is made when oxygen changes a fuel?
2. What happens when oxygen changes iron?
3. Why do you need food and oxygen?
4. What three things do you need to start a fire?

Chapter Review

Science Words: Unscramble the letters to find the correct terms.

1. Salt will EDILOSSV in water.
2. Sugar in water forms a ILNOOSTU.
3. AEHT helps solids to dissolve in liquids.
4. When a solution evaporates, the solids that form are ACLRSSTY.
5. EGNOXY is a gas in the air.

Questions: Answer in complete sentences.

1. How does heat help solids to dissolve in water?
2. Why do we say that the ocean is a solution?
3. How is food in your body like fuel?
4. Doug was in trouble. Yesterday, he was building a tree house. He left his father's hammer outdoors. How was the hammer different the next day? Why was it different?
5. If a solution of sugar and water evaporates, what is left?
6. Why is lemonade a solution?
7. Why does your body need oxygen?
8. What are three ways to put out a fire?
9. What is the name for the solids that are left after fuel burns?
10. What causes iron nails to rust?
11. What is oxygen?

INVESTIGATING

What happens to air when it is heated or cooled?

A. Gather these materials: 1-liter jar, plastic sandwich bag, rubber band, 2 bowls, hot water, cold water, and ice.

B. Squeeze all the air out of the plastic bag. Then, using a rubber band, fasten the bag over the top of the jar. Look at the picture below.

C. Set the jar in a bowl of hot water. Watch what happens.

1. What is in the jar?
2. What happened to the plastic bag?
3. Explain why this happened.

D. Now, set the jar in a bowl of ice water. Watch what happens.

4. What happened to the plastic bag?
5. Explain why this happened.
6. What do you think will happen if you put the jar back into the hot water?

E. Now, put the jar back into the hot water.

7. What happened to the plastic bag?

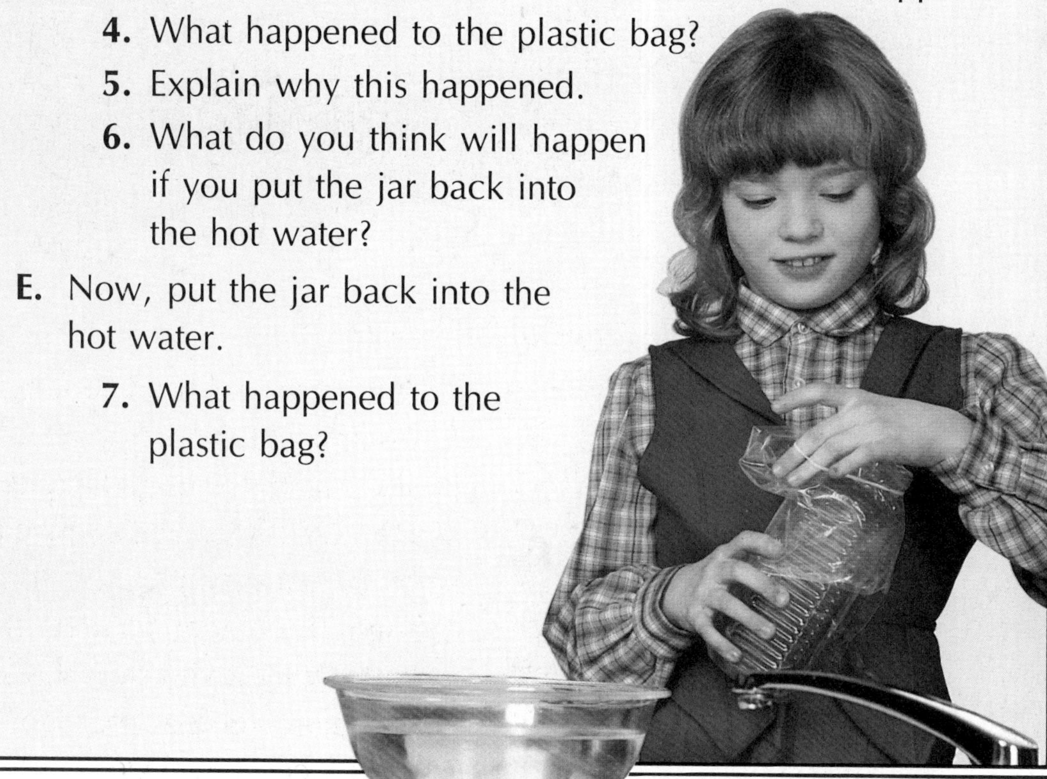

CAREERS

HEATING TECHNICIAN ▶

You have learned how heat changes matter. A heating technician (tek-**nih**-shun) plans and repairs equipment that controls temperatures in buildings. Technicians may help scientists plan and test new equipment. Some training is needed for a job as a technician.

◀ PHYSICIST

You have learned about the forms of matter. A physicist (**fiz**-ih-sist) studies matter, energy, space, and time. Some physicists study one area of physics, such as changes in matter. Other physicists study light and sound. Physicists have studied mathematics and science in college.

OUR SOLAR SYSTEM

UNIT 4

CHAPTER 10

THE EARTH AND THE MOON

10-1. The Blue Planet

Pretend you are in a spacecraft. You are traveling toward the earth. When you look out the window of the spacecraft, you see a blue planet. Do you know why the earth looks blue?

When you finish this section, you should be able to:

- [ ] **A.** Describe the *planet* earth.
- [ ] **B.** Tell what a *planet* is.
- [ ] **C.** Explain how the earth moves.

The earth is like a giant ball. It is very large and round. Look at the picture below. The top and bottom of the earth are white. The white material is ice. On the top and bottom of the earth are the **polar ice caps**. The darker colored places are land. What are the white swirls? They are clouds. The blue of the oceans and the layer of air around the earth make it look blue from space. Now you know why the earth is called the blue **planet**. A *planet* is a solid body in space that does not give off its own light.

Polar ice caps: Ice and snow that cover land near the top and bottom of the earth.

Planet: A solid body in space that does not give off its own light.

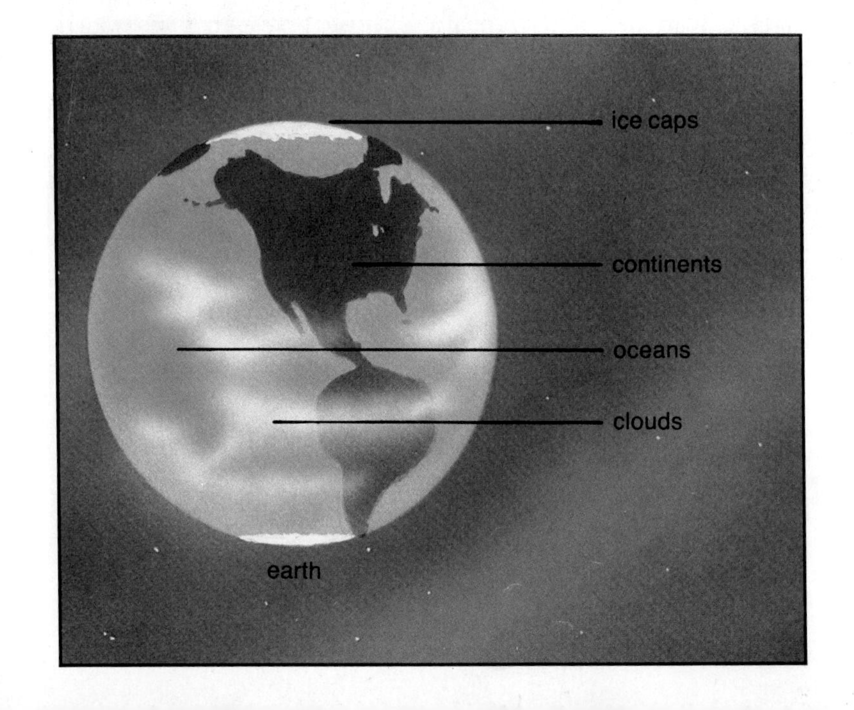

Revolution: The movement of one object around another object.

The planet earth moves around the sun once each year. It takes 365 days for the earth to travel around the sun. Each trip around the sun is called a **revolution** (rev-voh-**loo**-shun). A *revolution* is the movement of one object around another object. How many revolutions has the earth made around the sun since you were born? If you are eight years old, the earth has made eight revolutions.

The sun warms the earth. Even though the sun is far away, the heat from the sun keeps the earth warm. The part of the earth that is tilted nearest the sun will be the warmest. This is why we have seasons. We have summer when the part of the earth where we live is tilted toward the sun. When it is farther away from the sun, we have winter. What season is it now where you live? Is the earth tilted toward or away from the sun? How do you know?

Look at the picture on page 148. The earth has another movement. It spins on its **axis** (**ak**-sis) once each day. Think of the earth's *axis* as an imaginary line running from the North Pole to the South Pole. The North Pole is the point farthest north on the earth. The South Pole is the point farthest south. The spinning of the earth on its axis is called **rotation** (roh-**tay**-shun). The earth makes one *rotation* in 24 hours. This boy is spinning a basketball on his finger. The earth spins on its axis like the basketball. The spinning causes night and day. Let's see how.

Axis: An imaginary line running through the earth from the North Pole to the South Pole.

Rotation: The spinning of the earth on its axis.

The earth gets its light from the sun. Only the side facing the sun receives light. It is daytime on the side receiving light. The other side is dark. It is night there. As the earth turns, daylight slowly comes back to that part of the earth.

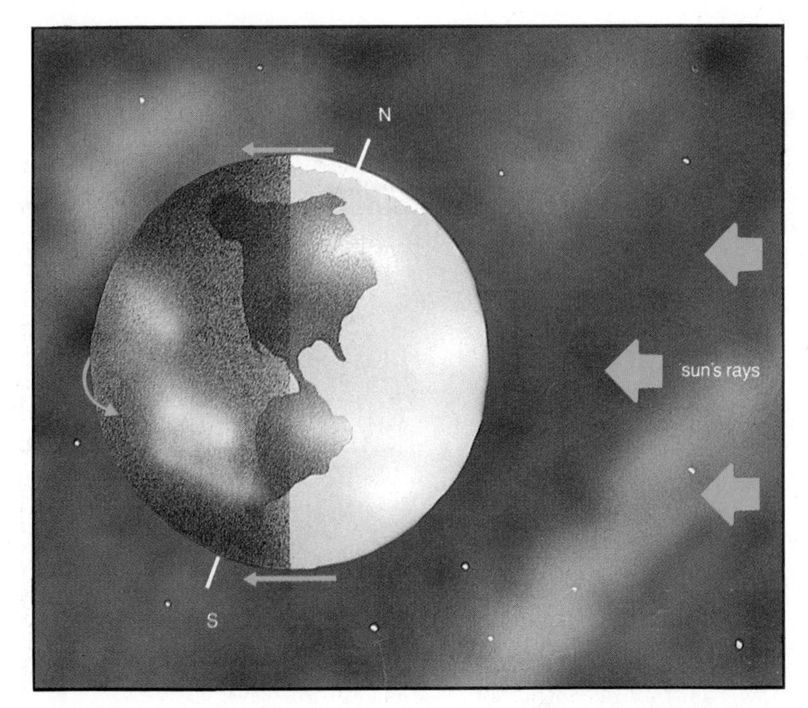

ACTIVITY

How does the earth's rotation cause night and day?

A. Gather these materials: chalk, globe, and flashlight.

B. Using chalk, mark your state on the globe.

C. In a dark room, shine the light at the globe.

1. How much of the globe is in light?
2. How much of the globe is in darkness?

D. While the globe is turned around once, watch your state.

3. When did daylight begin in your state?
4. When did darkness begin in your state?

Section Review

Main Ideas: The earth is a planet. A planet is a solid body in space that does not give off its own light. The earth revolves around the sun once each year. It makes one rotation on its axis every 24 hours.

Questions: Answer in complete sentences.

1. Why is the earth called the blue planet?
2. Name four things you would see if you were in space looking at the earth.
3. What keeps the earth warm?
4. What are two ways the earth moves?

10-2. Moon Watch

This person is trying to find out how the moon changes. She has been drawing pictures of the moon for many days. What do you think she will find out?

When you finish this section, you should be able to:

- ☐ **A.** Tell what a *satellite* is.
- ☐ **B.** Explain why the moon seems to change shape.

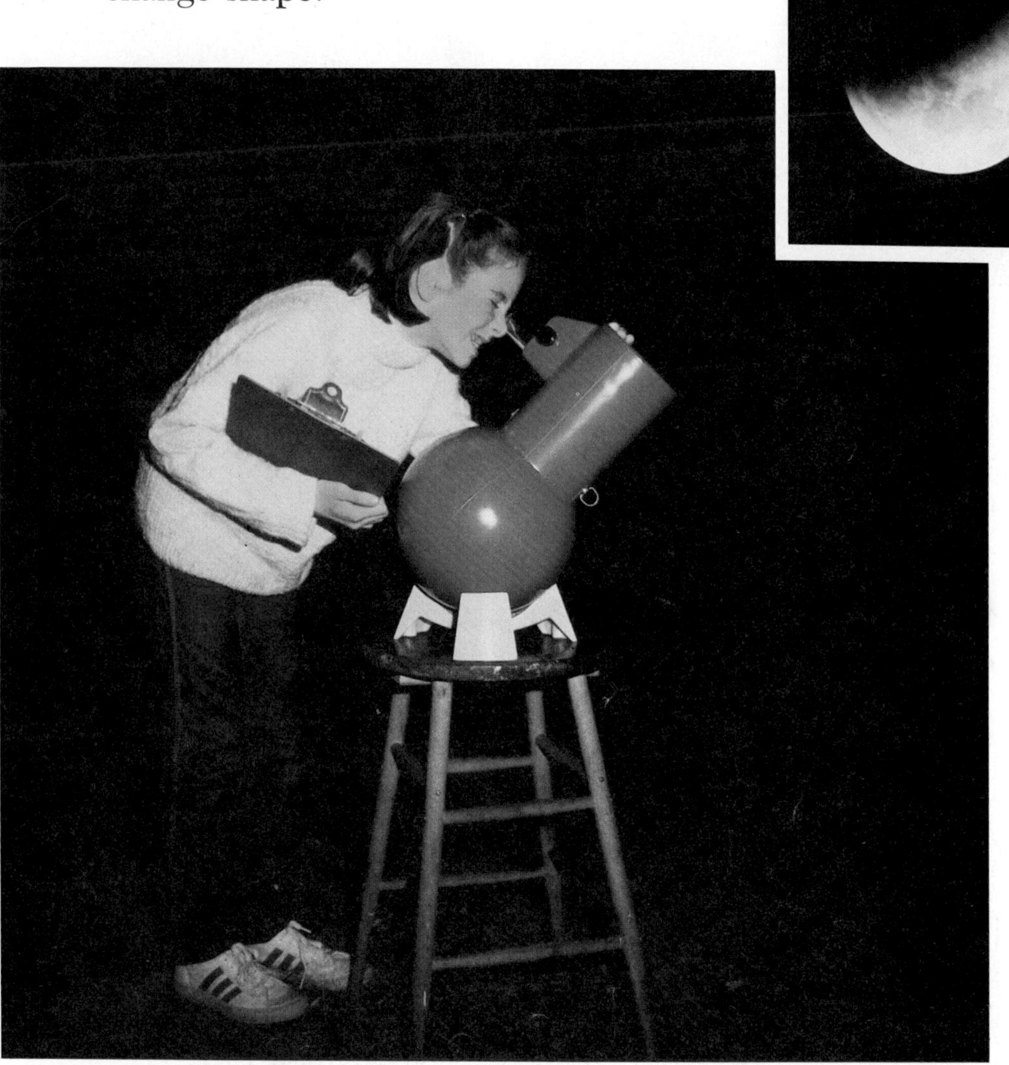

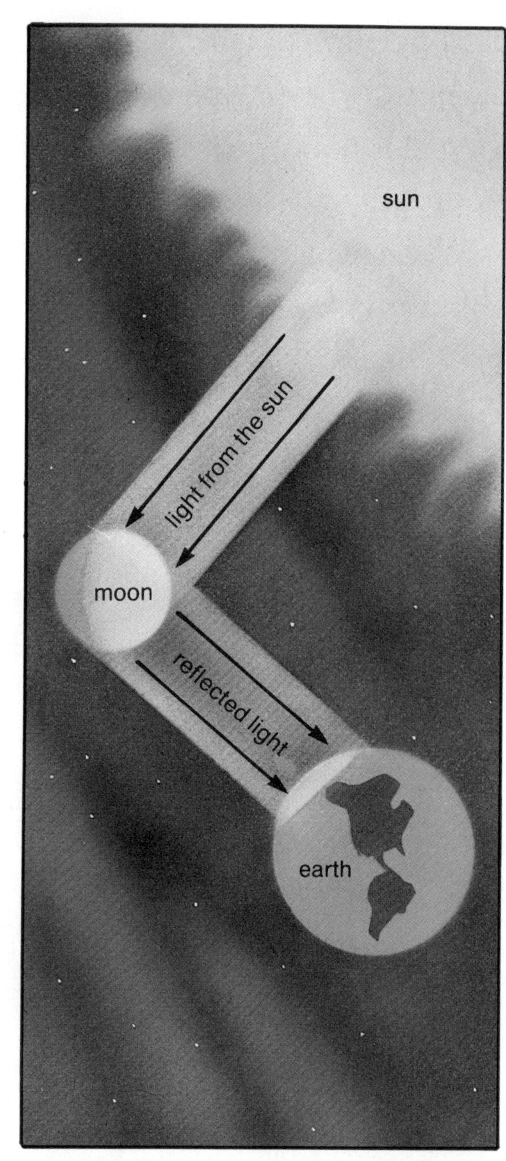

Satellite: A small planet that revolves around another planet.

Orbit: A path around an object in space.

Reflect: To bounce off.

The moon is a natural **satellite** (**sa**-teh-lite) of the earth. A *satellite* is a small planet moving around another planet. The picture shows the moon's **orbit**, or path, around the earth.

The moon does not give off its own light. We see the moon when sunlight shines on it. The sunlight **reflects**, or bounces off, the moon's surface.

Let's look at how the moon moves around the earth. One revolution of the moon takes about 28 days. Now, look at the picture. Pretend the boy is the earth. The girl is the moon.

Compare the moon's revolution around the earth to the earth's revolution around the sun. Which revolution takes longer? The earth's revolution around the sun is longer. It takes 365 days. The moon's revolution around the earth takes only about 28 days.

Do you ever look at the moon? Does it always have the same shape? No, it sometimes looks like a whole circle. At other times, we see only half of it from the earth. At still other times, we see only a small piece of it. The picture at the top of page 155 helps explain why.

ACTIVITY

What causes the moon to look different?

A. Gather these materials: globe, ball, and flashlight.

B. You will need 3 people for this activity. The person who holds the flashlight will be the sun. The person who holds the globe will be the earth. The person who holds the ball will be the moon.

C. Set up the sun, moon, and earth as shown in the picture. Have the sun shine light at the earth. Slowly move the moon around the earth. Stop at the places shown in the picture on page 155.

1. Pretend you are on the earth. How much of the moon can you see at each stop?
2. You see a full moon. Is the sun between you and the moon?

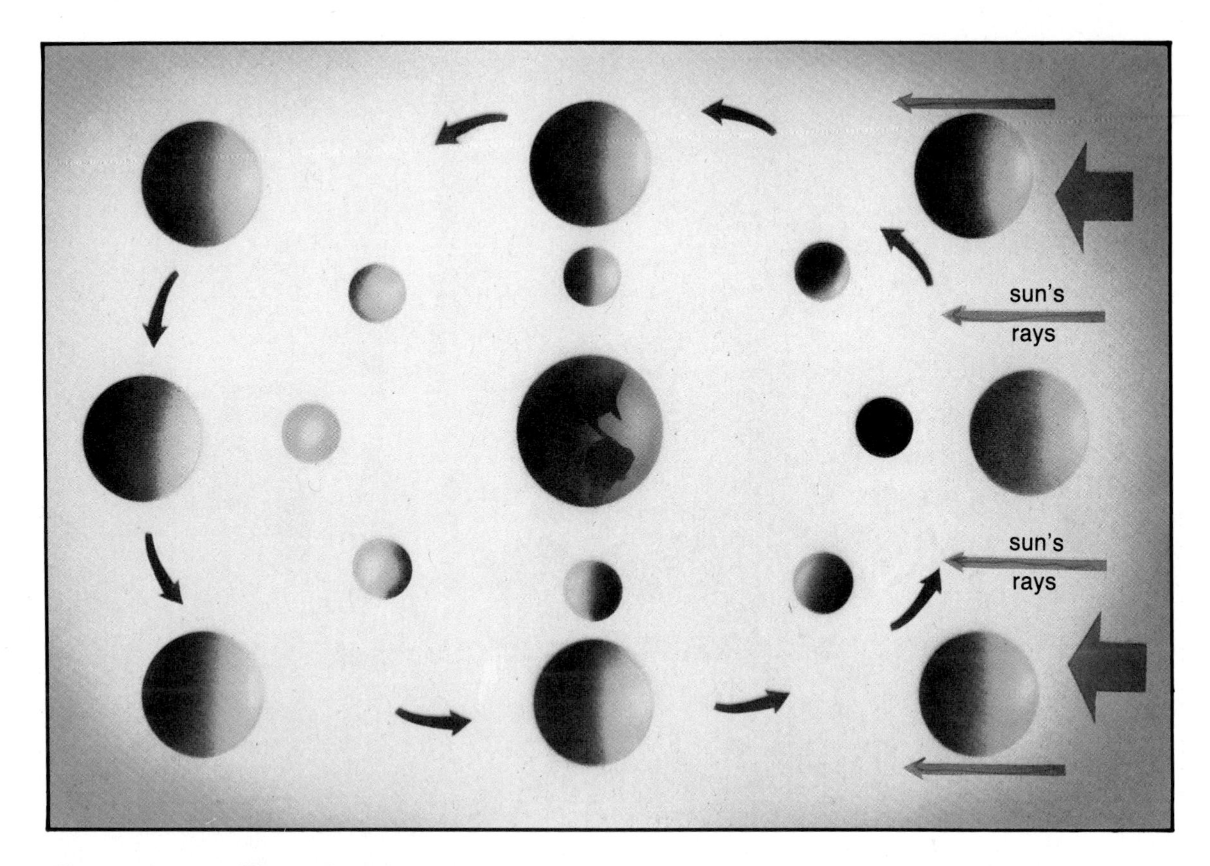

Section Review

Main Ideas: The moon is a satellite of the earth. It moves around the earth in an orbit. The moon seems to change shape as it moves around the earth.

Questions: Answer in complete sentences.

1. How long does it take for the moon to revolve around the earth?
2. Why is the moon called a satellite?
3. Does the moon make its own light? Explain your answer.
4. When you look at the moon from the earth, does it always have the same shape?

10-3.

Exploring the Moon

Who were the first two people to walk on the moon? In 1969, Neil Armstrong and Buzz Aldrin stepped from *Apollo 11* onto the surface of the moon. What do you think it would be like to walk on the moon?

When you finish this section, you should be able to:

- ☐ **A.** Compare the earth and the moon.
- ☐ **B.** Explain how moon *craters* are made.
- ☐ **C.** Describe the surface of the moon.

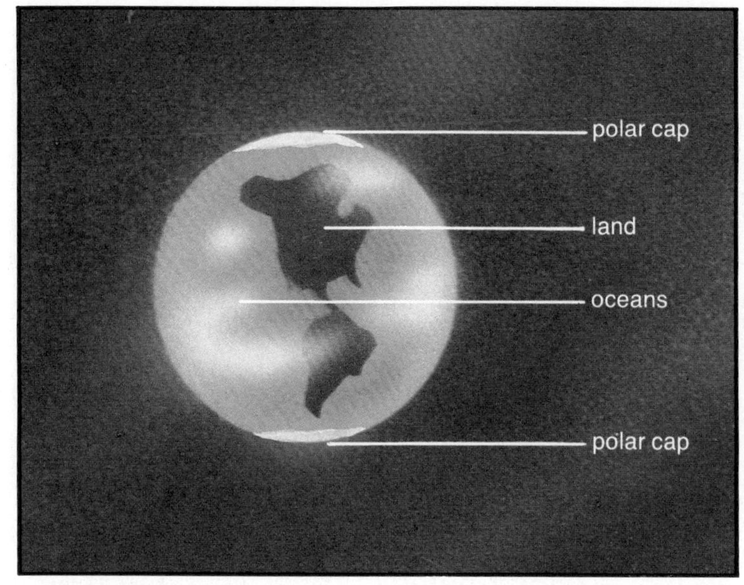

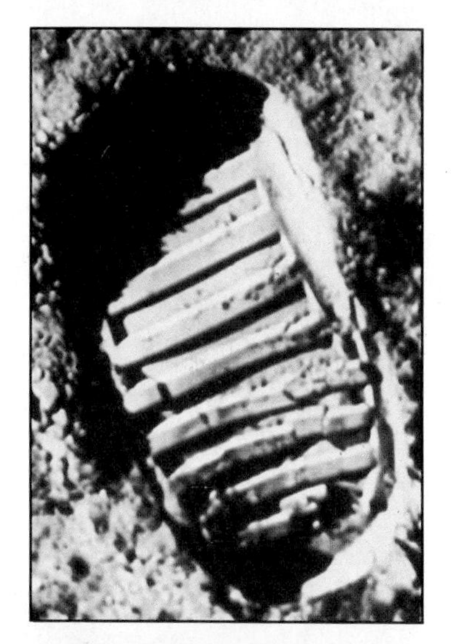

The moon is much smaller than the earth. Look at the picture on the left above. It shows the earth. Notice the polar ice caps, oceans, land, and clouds. Now look at the picture on the right above. It shows the moon. The moon is very different from the earth. The moon has no air. Travelers to the moon must bring their own air. There is no water on the moon. Because of this, there are no ice caps or oceans.

The surface of the moon is covered with a fine dust. How do scientists know this? Because of the footprints made by the astronauts (**as**-troh-nauts) when they landed on the moon. Such footprints could only be made in sandy or dust-like material.

There are many rocks on the moon. Some of the rocks are very big. When the astronauts returned from the moon, they brought some small moon rocks back.

If we look at the moon from a distance, we can see bowl-shaped holes of different sizes. These holes are called **craters** (**kray**-terz). Scientists think the *craters* were made when big pieces of stone and metal traveling in space hit the moon. Some craters are so large that they have smaller craters inside them. Look at the craters in the picture above.

Craters: Bowl-shaped holes made by objects traveling in space that hit the moon or the earth.

The earth also has craters. The picture below shows a large crater in Arizona. It was probably made a long time ago when an object traveling in space hit the earth.

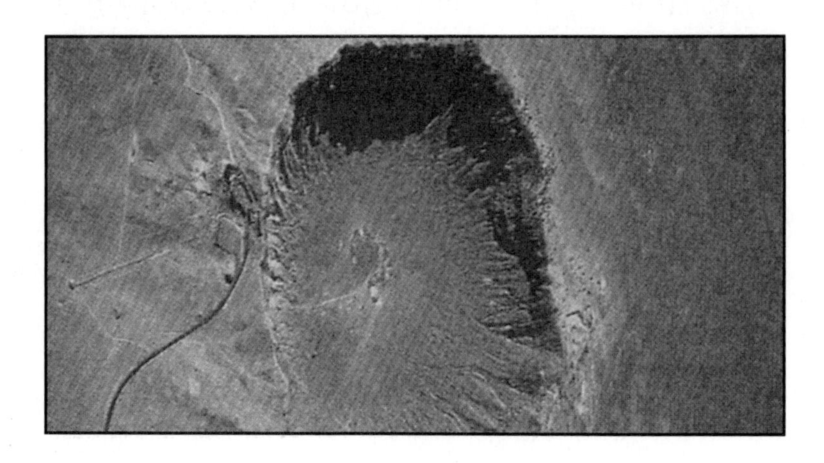

As shown in the picture at the right, there are also dark areas on the moon. When you look at them with a telescope, they look very flat. What do you think these dark spots are? A long time ago, scientists thought these dark areas were oceans. We now know they are rocks and dust.

ACTIVITY

How are craters formed?

A. Gather these materials: small tray, silt, tape, talcum powder, marble, small rock, and string.

B. Pour the silt into the tray. It should be 3 centimeters deep. Sprinkle a thin layer of talcum powder over the silt.

C. Tape the end of a piece of string to the rock. Then hold the rock about 30 centimeters above the tray. Drop the rock into the silt. Carefully lift the string to remove the rock.

1. What happened to the silt when the rock hit it?
2. What shape was formed in the silt?

D. Using the marble instead of the rock, repeat step C.

3. What was the shape of the crater?
4. How do you think craters on the moon are formed?

Section Review

Main Ideas: The moon is smaller than the earth. The moon's surface is covered with dust and rocks. The craters on the moon were made when objects traveling in space hit the moon.

Questions: Answer in complete sentences.

1. How is the moon different from the earth?
2. What are the dark areas on the moon?
3. How are moon craters made?
4. What must travelers to the moon take with them?

Carl Sagan

Dr. Carl Sagan (**Say**-gun) works at the Laboratory for Planetary Studies at Cornell University. Dr. Sagan studies the planets and stars.

An area of special interest to Dr. Sagan is the study of Venus and Mars. His first public report on Mars was made in 1956. At that time, Mars was closer to the earth than it had been in 32 years.

Dr. Sagan has worked on America's space program. In 1970, he received NASA's Apollo Achievement Award. Dr. Sagan has written many books about the planets and stars. One of his books was *Cosmos*. A TV program was based on this book.

CHAPTER REVIEW

Science Words: List the letters **a** through **g** on paper. Write the correct word from the list next to each letter.

rotates
craters
orbit
polar ice caps
revolves
reflected
satellite

Ice and snow that cover part of the earth are called ___a___. The earth ___b___ around the sun once every 365 days. The earth ___c___ once every 24 hours.

The moon is a ___d___ of the earth. The moon travels in a path, called an ___e___, around the earth. The light we see from the moon is sunlight ___f___ from the surface of the moon. ___g___ are bowl-shaped holes in the moon's surface.

Questions: Answer in complete sentences.

1. Pretend you are on the moon. What would the earth look like from there?
2. How long does it take for the earth to revolve around the sun?
3. What is a planet?
4. What is a satellite?
5. How do we know the moon is covered with a dust-like material?

CHAPTER 11

The Sun and the Stars

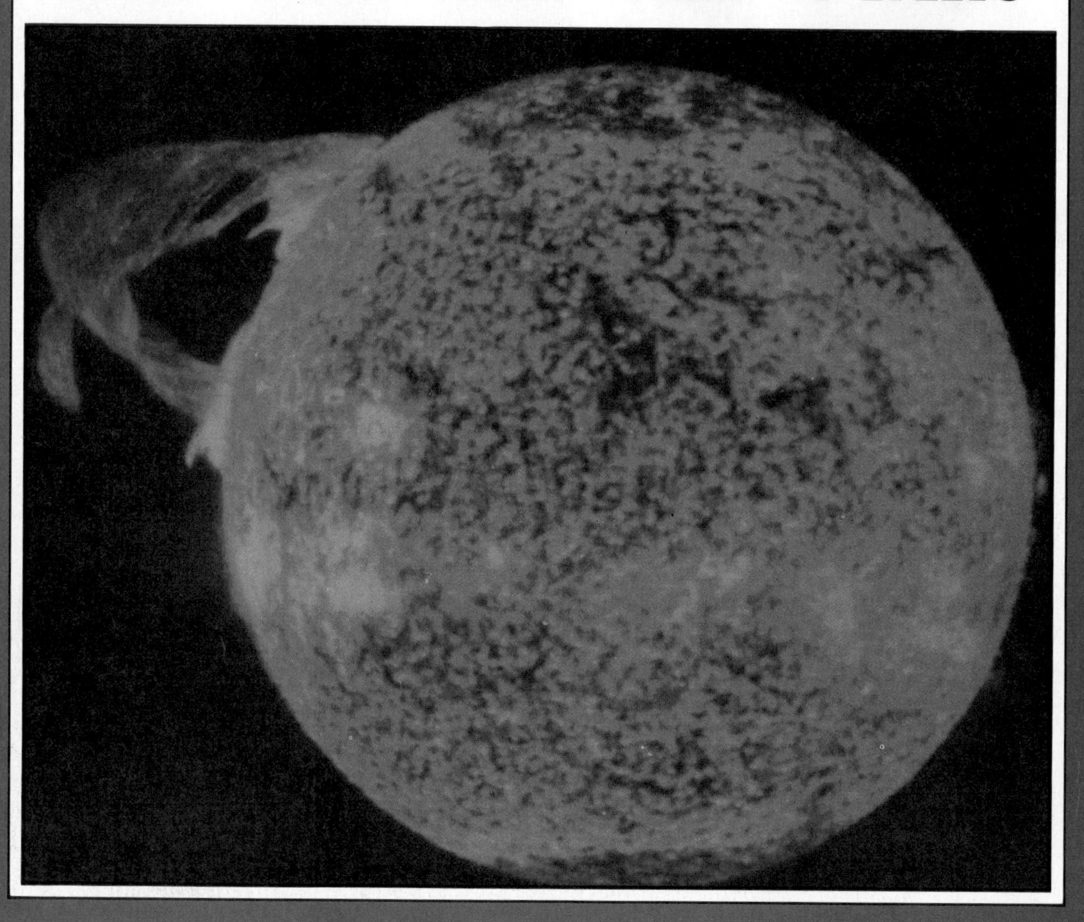

11-1. The Sun

This is a picture of the star closest to the earth. What is this star? It is the sun. If you could travel to the sun, what do you think you would find there?

When you finish this section, you should be able to:

- ▢ **A.** Compare the sun and the earth.
- ▢ **B.** Name some properties of the sun.

The sun is very large. To get an idea of how large the sun is, look at the drawing. The small circle is the earth. The sun is so big when compared to the earth that it will not fit on the page.

The sun is different from the earth in another way. The earth is a planet. The sun is a **star**. A *star* is made of hot gases. It is shaped like a ball. Do you know how the sun's light is made? The gases of the sun are so hot they give off heat and light. Since the earth is a planet, it does not give off light.

Star: A ball made of hot gases.

Let's take a closer look at the sun. Sometimes its gases appear to be leaping high off the surface. This movement tells scientists that the sun's surface is very hot and active. Do you see the dark spots in the picture on page 162? These dark spots are **sunspots**. *Sunspots* are places where the gases have cooled.

Sunspots: Dark spots on the sun's surface.

ACTIVITY

How does the sun's gravity affect the earth?

A. Gather these materials: string (60 centimeters long), tape, empty spool of thread, and Styrofoam ball.

B. Tape the string to the Styrofoam ball.

C. Slide the string through the empty spool. Hold the spool with one hand. Hold the end of the string with your other hand. Whirl the ball around your head.

1. What path does the ball take as it is whirled around?

D. Whirl the ball again. This time feel how the string is being pulled.

2. What is causing the string to be pulled?
3. What would happen to the ball if you cut the string while whirling the ball?
4. How is the string like the sun's gravity?

The earth travels in an orbit around the sun. Do you know why the earth does not leave this orbit? The sun pulls on the earth. This pull is called **gravity**. The *gravity* of the sun keeps the earth in its orbit.

Gravity: The pull of the sun on the earth.

Section Review

Main Ideas: The sun is a star. The sun is very large and very hot. The sun's pull keeps the earth in orbit.

Questions: Answer in complete sentences.

1. What are two ways the sun is different from the earth?
2. How is a star different from a planet?
3. What is a sunspot?
4. What keeps the earth orbiting around the sun?

11-2. The Night Sky

Look at the picture below. Do you know what it is? It is a meteor (**mee**-tee-ohr). Have you ever seen streaks of light moving quickly across the sky? Those streaks of light were meteors.

When you finish this section, you should be able to:

- **A.** Identify objects in the night sky.
- **B.** Name four *constellations*.
- **C.** Explain how to find the *North Star*.

Meteor: A piece of stone or metal from outer space.

A **meteor** is made of stone or metal. It comes from outer space. A *meteor* is pulled into the air around the earth. It travels so fast that it gets very hot. The oxygen in the air makes it burn. It

gives off heat and light. Sometimes meteors burn up before they reach the ground. But some meteors land on the earth. Then, they are called **meteorites**. Do you know what caused the hole in the ceiling of this house? It was a *meteorite* that weighed 1 1/2 kilograms (3 pounds).

Meteorite: A stone or metal object that fell to earth from space.

Most objects in the night sky are stars. On a clear night, you can see millions of stars. If you look carefully, you will see that some stars are brighter than others. Do you know why those stars are brighter? One reason is that some stars are bigger than others. The larger a star is, the brighter it appears to us. Look at this picture. It shows how the sun compares in size to some other stars.

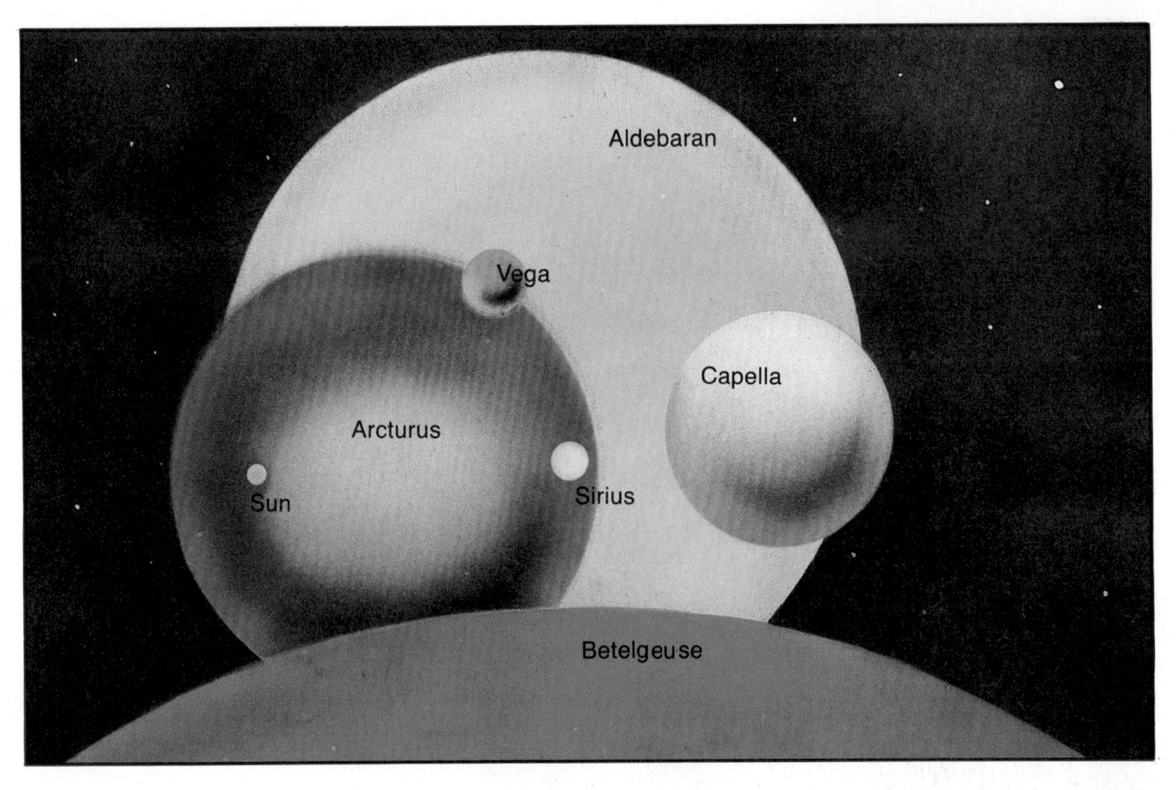

Another reason some stars look brighter is that they are closer to the earth. Stars farther away do not look as bright.

People have always liked to look at stars. A long time ago, people gave names to the groups of stars that make shapes in the sky. Have you ever seen these shapes in the night sky? These groups of stars are called **constellations** (kahn-steh-**lay**-shunz).

Constellation: A group of stars.

Look at the pictures above. What do you see? This *constellation* is called Draco (**dray**-koh), or the dragon. It is a group of 16 stars. Have you ever seen this constellation? To find the dragon in the sky, connect the stars with straight lines. In the picture on the right, a dragon has been drawn around the stars in this constellation.

There are 88 constellations in the sky. Some are very hard to find. But let's look at three constellations that are easy to see.

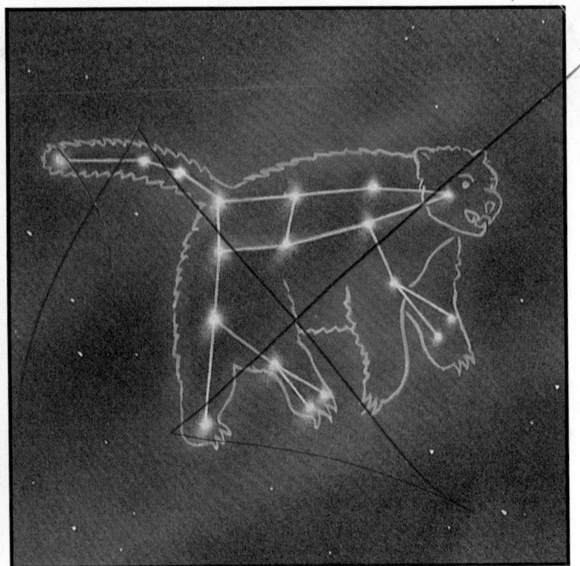

What animal do you see in the pictures above? It is the Great Bear. There are many stars in this constellation. Do you think the shape on the left looks like a bear? The story about this constellation was passed on to us by American Indians. In this tale, three brothers saw the bear go into a cave one night. The bear knew the brothers were near. So it put a net, which could not be seen, outside the cave. While the bear was sleeping, the brothers tried to get into the cave. But they were trapped by the net.

There is another constellation within the Great Bear. Look at the picture at the right. Can you find the smaller constellation? It is called the Big Dipper. It looks like a big spoon, or dipper. It is a very important constellation. If you can find the Big Dipper, you will be able to find the **North Star**. The *North Star* is always in the northern sky.

North Star: A star that is always in the northern sky.

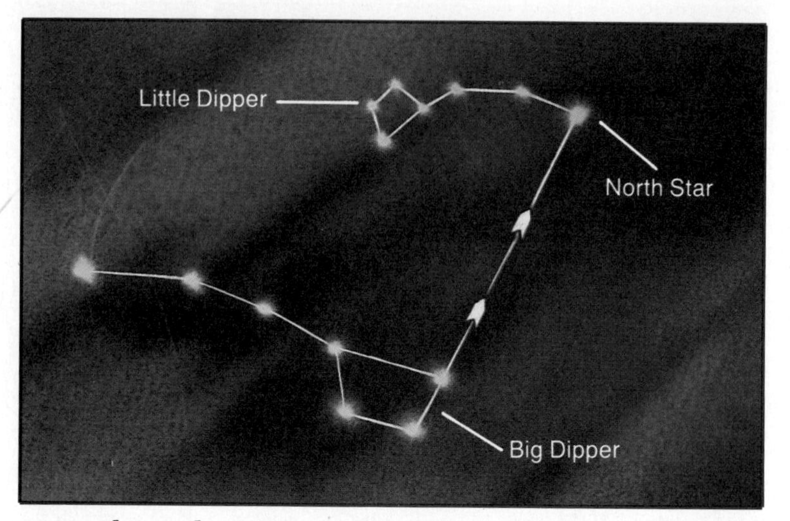

Look at the star map above. It shows the Big Dipper and another constellation. What do you think this constellation is called? It is called the Little Dipper. The last star on the handle of the Little Dipper is the North Star. Notice the broken line in the picture. It shows you how to find the North Star when you know where the Big Dipper is.

The star map below shows these three constellations. On the next clear night, use this map to find the constellations.

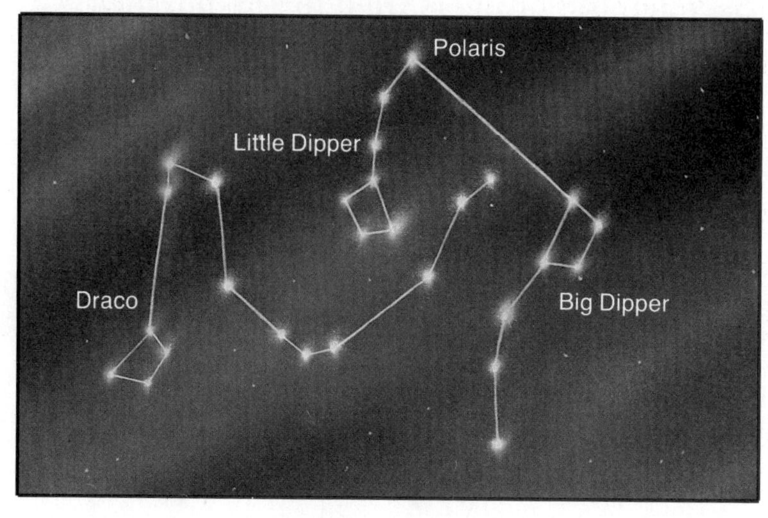

ACTIVITY

What are constellations?

A. Gather these materials: empty tin can, tape, flashlight, and black construction paper.

B. Draw the Big Dipper, the Little Dipper, and Draco on the paper. Using a pencil, make small holes where the stars are.

C. Tape the paper onto one end of the can. Darken the room. Aim the can toward the wall. Shine the light into its open end.

1. What constellations do you see?
2. When the Big Dipper is shown on the wall, where should the North Star be?

Section Review

Main Ideas: Meteors and stars are objects in the night sky. Constellations are groups of stars. The North Star can be found by finding the Big Dipper.

Questions: Answer in complete sentences.

1. What is a meteorite?
2. What is a constellation?
3. Make a drawing of one constellation. Label your drawing.
4. Draw a map that shows how to find the North Star.

CHAPTER REVIEW

Science Words: Match the terms in column A with the definitions in column B.

Column A	Column B
a. North Star	1. A ball of hot gases
b. Star	2. Dark spots on the sun's surface
c. Gravity	3. The pull of the sun on the earth
d. Sunspots	4. A streak of light seen in the night sky
e. Big Dipper	5. A stone or metal object that falls to earth from space
f. Meteor	6. A group of stars
g. Meteorite	7. A constellation that looks like a dragon
h. Draco	8. A constellation that looks like a big spoon
i. Constellation	9. A star that is always in the north

Questions: Answer in complete sentences.

1. What are the dark spots on the sun?
2. Why doesn't the earth give off light?
3. What is a meteor?
4. Why do some stars look brighter than others?
5. How many constellations are there in the sky?
6. Describe the sun.

CHAPTER 12

JOURNEY TO THE PLANETS

12-1. The Solar System

This is a picture of the *Voyager* spacecraft. It left the earth in 1977. It was going to the edge of our solar system. How long do you think it took to get there?

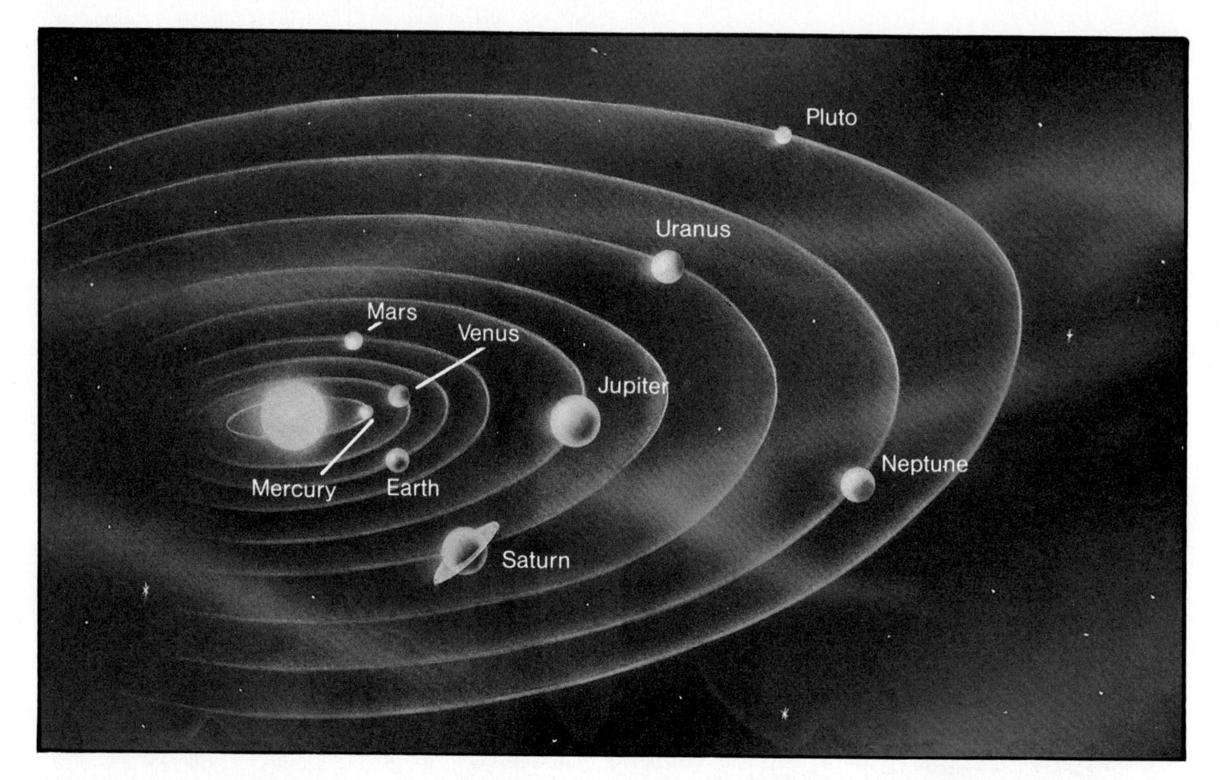

When you finish this section, you should be able to:

- **A.** Identify the objects in the *solar system.*
- **B.** Describe how the planets move around the sun.
- **C.** Compare the sizes of the planets.

Solar system: The nine planets, their moons, and the sun.

The nine planets, their moons, and the sun are called the **solar system**. The drawing above shows the *solar system.* The sun is in the center. It is the largest object in the solar system.

The planets revolve around the sun in orbits. Each planet travels in its own orbit. It takes only 88 days for Mercury to revolve around the sun.

But it takes 248 years for Pluto to make one orbit around the sun.

Look at the picture below. The planets are numbered from one to nine. Beginning with number one, the planets are Mercury, Venus, Earth, Mars, Jupiter, Saturn, Uranus, Neptune, and Pluto. Mercury is nearest to the sun. Pluto is the planet farthest from the sun.

There are two groups of planets in the solar system. In one group, there are earth-like planets. These are the inner planets. Look at the picture. Compare the sizes of these planets.

The planets in the second group, except Pluto, are very large. These planets are the outer planets. The outer planets have many more moons than the inner planets.

ACTIVITY

How do the planets compare in size?

A. Gather these materials: pieces of clay of different colors, basketball, and newspapers.
B. Make a small model of the solar system. The picture on page 175 shows the sizes of the planets. Make a ball of clay for Mercury. Put it on a sheet of newspaper. Label it.
C. Repeat step B for each planet.
D. Use the basketball for the sun.
  1. What is the largest planet?
  2. What is the smallest planet?

Section Review

Main Ideas: The solar system is made up of the sun, the nine planets, and their moons. The sun is the largest object in the solar system. The planets revolve around the sun in orbits. The inner planets are smaller than the outer planets.

Questions: Answer in complete sentences.

1. Draw a picture of a planet orbiting the sun.
2. What is in the center of our solar system?
3. How long does it take for Mercury to revolve around the sun?
4. Name the nine planets in our solar system.

12-2. The Earth-like Planets

Pretend you are aboard the spacecraft shown here. You are about to take a trip to the planets. First, you will visit planets close to the earth. These planets are about the same size. What do you think you will find on these planets?

When you finish this section, you should be able to:

- **A.** Describe the main features of Mercury, Venus, and Mars.
- **B.** Compare these three planets to the earth.

The path of your trip is shown in the margin. You will first visit Mercury. Then, you will go to Venus. Mars will be the last planet on the trip.

Until 1974, scientists did not know what Mercury looked like. The picture below shows the surface of Mercury. It looks like our moon. How do you think the craters on Mercury were made?

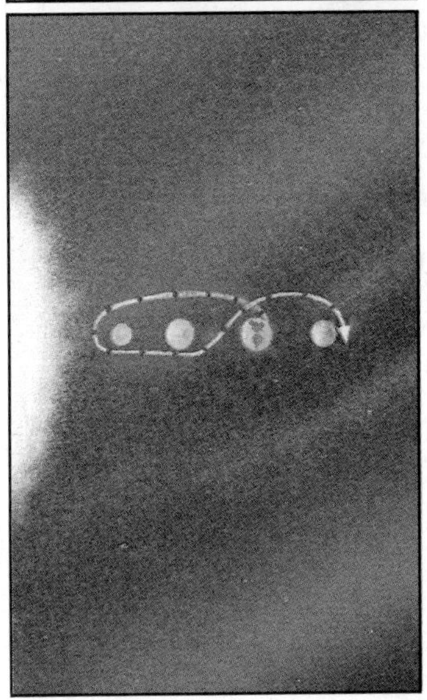

Mercury is the planet closest to the sun. It orbits the sun once every 88 days. On the side facing the sun, Mercury's surface temperature rises to over 400°C (752°F). On the dark side, which does not face the sun, it is very cold. The temperature on the dark side falls to about −200°C (−328°F).

Let's take a close look at Mercury. As you can see, it would be difficult to land on Mercury. Its surface is covered with craters, hills, plains, and cliffs. Mercury probably never had air. Meteorites could easily crash on the surface without burning up. Some scientists believe this is how the craters were formed. Mercury's surface has probably changed very little during the past 3 billion years.

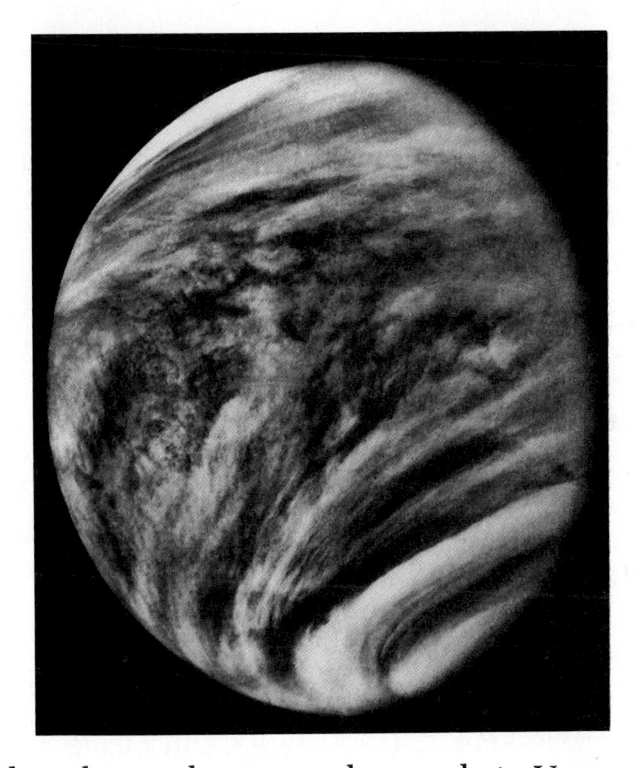

The planet closest to the earth is Venus. As shown in the picture above, Venus is almost always covered with thick clouds. So, it is very hard to see the surface of Venus. Venus receives about twice as much sunlight as the earth. This sunlight is reflected by the thick cloud cover. This reflection makes Venus one of the brightest objects in the sky. Sometimes you can see it as the evening or morning "star."

For a long time, scientists thought Venus was very much like the earth. But they have found that this is not so. The surface of Venus is very hot, with a temperature of 475°C (887°F). Its thick cloud cover is made up almost entirely of the gas *carbon dioxide*. This gas traps heat. Thus, the surface of Venus is like a furnace.

Venus turns very slowly on its axis. So, days and nights on that planet are very long.

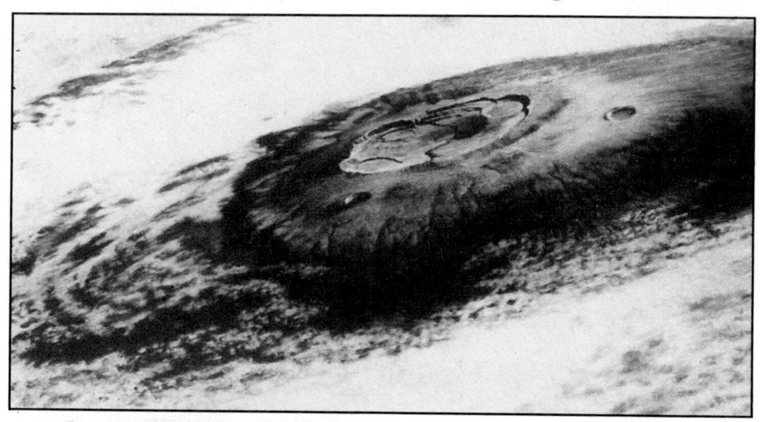

The next planet on our trip is Mars. The picture above shows Mars as we come close to it. The bright red dot is a large volcano on the planet. Do you know why Mars is called the red planet? It is because the soil is reddish brown.

If you could look up through the air of Mars, it would seem pink. It would look like the color of the sky during some sunsets on earth.

In 1976, two *Viking* spacecrafts landed on Mars. Scientists wanted to know if there was life on that planet. Do you know what they learned? Their tests showed there are no living things on Mars.

Like the earth, Mars has polar ice caps. The air on Mars is very thin. And it is very cold there. In winter, temperatures can drop to −130°C (−202°F). There are strong winds on Mars during the summer. These winds cause great dust storms that cover most of the planet. These storms sometimes last for several months.

ACTIVITY

Making a mission report.

A. Your teacher will give you a mission report to complete.

B. Your mission will begin with the planet Mercury. Reread this section. Then, read other books about the planets. Look at pictures of the planets. Complete the mission report.

Section Review

Main Ideas: Mercury, Venus, and Mars are earth-like planets. They are close to the sun. Mercury, the closest planet, is covered with craters. The hottest planet is Venus. It receives about twice as much sunlight as the earth. It is covered with clouds. Mars is called the red planet because its soil is reddish brown.

Questions: Answer in complete sentences.

1. What is the temperature on the side of Mercury facing the sun?
2. Which planet is closest to the earth?
3. Why is Mars called the red planet?
4. Which planet is one of the brightest objects in the sky?
5. What is the temperature on Mars during the winter?

12-3. The Giant Planets

Jupiter is a very large planet. The distance around it is 11 times that of the earth. Jupiter is made up of twice as much matter as all the other planets added together. What else do you think we will learn about the planets as we continue our trip?

When you finish this section, you should be able to:

- ☐ **A.** Describe the main features of Jupiter, Saturn, Uranus, and Neptune.
- ☐ **B.** Name the smallest planet.

Jupiter is the largest planet in the solar system. This planet has at least 16 moons. Jupiter is bigger than many stars. It rotates very

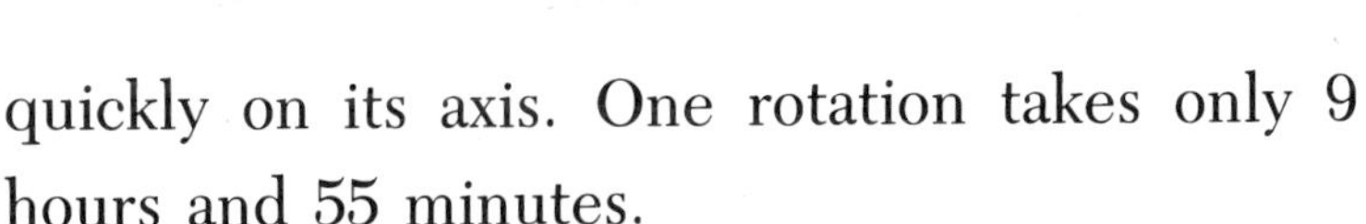

quickly on its axis. One rotation takes only 9 hours and 55 minutes.

Most of the air around Jupiter looks like swirls of clouds. But there is a Great Red Spot that always appears in the same place. Some scientists think this spot may be caused by a large storm. The Great Red Spot is shown above.

Jupiter's moons are as important as the planet itself. Four of Jupiter's moons are the size of small planets. Pictures of these moons are shown at the right. Callisto (kuh-**list**-oh) is covered with a thick, icy crust. There are many craters on this moon. The biggest moon, Ganymede (**gan**-uh-meed), also has many craters. Europa (yoo-**roh**-puh) has an icy crust. Scientists believe this crust may be only a few kilometers deep. A deep ocean may be under the crust. Io (**i**-oh) looks like a pizza. The dark spots on it are volcanoes.

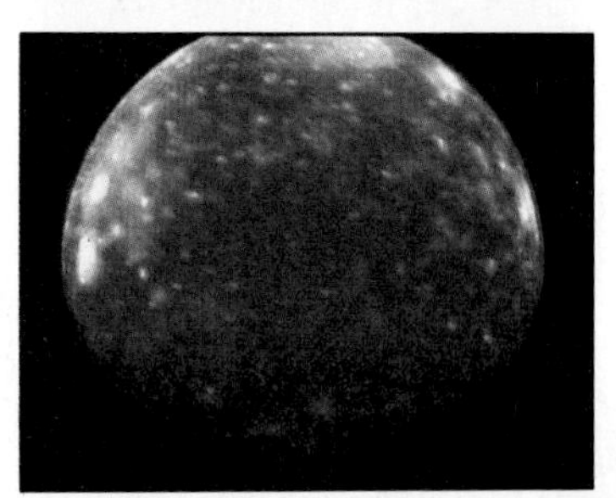

The next planet on our trip is Saturn. It is known for its beautiful rings. The rings circle Saturn. They are made of pieces of ice. Some of those pieces are as small as bits of dust. Other pieces are as big as a house.

Like Jupiter, Saturn rotates very quickly. It turns on its axis once every 10 hours and 30 minutes. But it takes almost 30 years for Saturn to make one revolution around the sun.

Saturn has at least 17 moons. One moon, called Titan (**ty**-tun), is between the size of Mercury and Mars. Most of Saturn's moons are very cold. But Titan is different. It has a thick **atmosphere** (**at**-mus-**feer**). The gases that surround a planet are called its *atmosphere*. As far as we know, Titan is the only moon in the solar system with an atmosphere.

Atmosphere: The gases that surround a planet.

We are now ready to visit the last three planets on our solar trip. We will travel to the edge of our solar system. Because these planets are so far away, there is still much to be learned about them.

Uranus is about four times larger than the earth. It is tilted so far on its axis that it appears to be lying on its side. Scientists are not sure how long Uranus takes to rotate on its axis. But it seems to take between 13 and 25 hours. It takes

ACTIVITY

How large is the solar system?

A. Gather these materials: meter stick, metric ruler, adding-machine tape, and modeling clay.

B. Make clay models of the 9 planets in the solar system.

C. Cut a piece of tape 4 meters long. Then, place the planets along the tape according to the chart below.

Planet	Distance from End of Tape
Mercury	4 cm
Venus	7 cm
Earth	10 cm
Mars	15 cm
Jupiter	52 cm
Saturn	95 cm
Uranus	192 cm
Neptune	300 cm
Pluto	395 cm

Uranus 84 years to go around the sun. Uranus has at least five moons and nine rings.

The next planet, Neptune, is a lot like Uranus. It takes 165 years for the planet to go around the sun. At least two moons orbit Neptune. One moon, called Triton (**try**-tun), is very large. In 1989, *Voyager* will fly by Neptune. Maybe then we will find out if Neptune has rings like Uranus.

The last planet on our trip is Pluto. It is the smallest planet. Some scientists think Pluto may have been a moon of Neptune at one time. It takes over 248 years for Pluto to go around the sun. It has one moon, named Charon (**kar**-un).

Section Review

Main Ideas: Jupiter is the largest planet in the solar system. Saturn is circled by rings. Uranus, Neptune, and Pluto are far from the earth.

Questions: Answer in complete sentences.

1. How long does it take for Neptune to go around the sun?
2. What is Jupiter's largest moon?
3. How is Titan different from other moons in the solar system?
4. How long does it take for Uranus to go around the sun?
5. Name the smallest planet.

CHAPTER REVIEW

Science Words: Write the sentences below on paper. Fill in the blanks with the correct words from the list.

Viking	**Venus**	**Mercury**	**Mars**
solar system	**Ganymede**	**Jupiter**	**Pluto**

1. ______ is the largest planet in the solar system.
2. Like the earth, ______ has polar ice caps.
3. ______ is the planet closest to the sun.
4. Our ______ is made up of nine planets, their moons, and the sun.
5. The ______ spacecraft landed on Mars in 1976.
6. ______, the largest moon of Jupiter, has many craters.
7. The smallest planet is ______.
8. ______ receives about twice as much sunlight as the earth.

Questions: Answer in complete sentences.

1. Which planet is called the red planet?
2. Name two of Jupiter's moons.
3. Name the three planets farthest from the sun.
4. Which two planets have rings around them?
5. What is the temperature on the surface of Venus?
6. How long does it take for Pluto to orbit the sun?

INVESTIGATING

How can the distance of a star from the horizon be measured?

A. Gather these materials: metric ruler, protractor, thumbtacks, string (10 centimeters long), and washer.

B. Make a copy of this chart.

Position of Star	First Angle	Second Angle
1.		
2.		
3.		
4.		

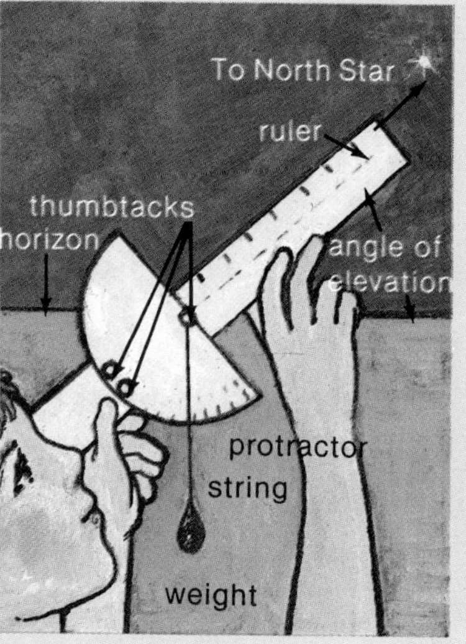

C. Attach the protractor to the ruler with 3 thumbtacks as shown.

D. Tie the string to the center tack so it hangs down. Attach the washer to the string. You have made an astrolabe (**as**-truh-layb), an instrument used to measure how far above the horizon an object is.

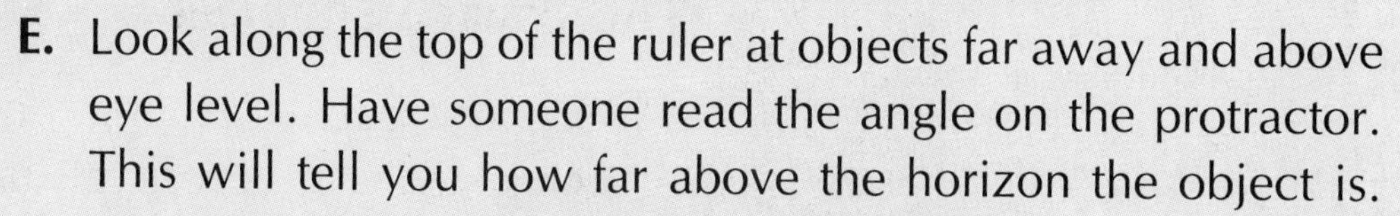

E. Look along the top of the ruler at objects far away and above eye level. Have someone read the angle on the protractor. This will tell you how far above the horizon the object is.

F. Choose 4 stars. Using your astrolabe, measure the angles for each star in your chart. Record your findings.

G. Repeat the measurements 1 hour later.

1. Did the angles of the stars above the horizon change?
2. If so, how did they change?

CAREERS

Astronomer ▶

You have learned about our solar system. An astronomer (as-**trah**-nuh-mer) studies the solar system and other objects in the sky. Using a telescope (tel-uh-skope), an astronomer can see far into space. TV cameras in spacecraft give them an even better look at planets. Astronomers study science in college.

◀ Aerospace Engineer

You have been studying about outer space. An aerospace (**ayr**-oh-spays) engineer designs and tests aircraft and spacecraft. Some engineers study the guiding of satellites. Others study the effect of spacecraft design on the environment. Aerospace engineers study engineering in college.

MAGNETISM AND ELECTRICITY

UNIT 5

CHAPTER 13

MAGNETS

13-1.

What Are Magnets?

These children are having fun. They are picking up objects without touching them. Have you ever done this? Do you know what the children are using?

When you finish this section, you should be able to:

☐ **A.** Tell what a *magnet* can do.

☐ **B.** Name things *magnets* can pick up.

The children in the picture on page 192 are using **magnets**. *Magnets* are objects that pick up or stick to certain things.

Magnets: Objects that pick up or stick to certain things.

A magnet is made up of a metal called **iron**. It can pick up a needle, a paper clip, or a thumbtack. A magnet cannot pick up chalk, coins, or rubber bands. A magnet can pick up or stick to objects made of *iron*. **Steel** is a metal made from iron. *Steel* can usually be picked up by a magnet.

Iron: A metal that can be picked up by a magnet.

Steel: A metal made from iron, which a magnet can usually pick up.

This child is using a magnet. It will stick to the refrigerator. Do you know why? The refrigerator is made of steel. So, the magnet holds the paper in place.

ACTIVITY

Can you guess what a magnet can pick up?

A. Gather these materials: chalk, coin, magnet, needle, paper, paper clip, pencil, ice-cream stick, rubber band, and thumbtack.

B. Place the materials, except for the magnet, on your desk.

  1. Which objects do you think the magnet can pick up?

C. Try to pick up the chalk and the other objects with the magnet.

  2. Which objects did the magnet pick up?

  3. Which objects were not picked up?

Magnets come in many shapes and sizes. The picture below shows many kinds of magnets. Find the magnet shaped like a doughnut. Can you find the horseshoe magnets in the picture?

There are many uses for magnets. Have you ever used a magnet? Maybe you have and didn't know it. Look at the pictures above. How is each child using a magnet?

Section Review

Main Ideas: Magnets can pick up or stick to iron and steel. Magnets come in many shapes and sizes. Magnets are put into some can openers, pot holders, and games.

Questions: Answer in complete sentences.

1. What is a magnet?
2. Name two metals a magnet can pick up or stick to.
3. Draw a picture of a horseshoe magnet.
4. Name an object a magnet will *not* pick up.
5. Which of these objects can be picked up by a magnet: chalk, a rubber band, a plastic button, a paper clip, paper, a nail?

13-2.

How Can We Use Magnets?

Lee has a problem. A box of pins fell onto the rug. How can he pick up the pins without hurting his fingers?

When you finish this section, you should be able to:

- ☐ **A.** Name the strongest parts of a magnet.
- ☐ **B.** Tell how the ends of magnets can affect one another.
- ☐ **C.** Explain how a magnet in a *compass* can tell direction.

Poles: The ends of a magnet.

Lee could use a magnet to pick up the pins on the rug. Most of the pins will stick to the ends of a magnet.

The ends of a magnet are called its **poles**. Magnets are strongest at their *poles*. That is why the ends of a magnet can pick up many pins. Magnets have a north pole and a south pole. Sometimes the letter N is written on the north pole of a magnet. What letter might be written on the south pole?

If you bring the north pole of a magnet near the south pole of another magnet, the poles pull on each other. A north pole and a south pole are **unlike poles**. *Unlike poles* pull on each other.

Unlike poles: Opposite poles; a north and a south pole.

If you bring the north pole of a magnet near the north pole of another magnet, something else will happen. The poles push away from each other. The same thing will happen if two south poles are brought near each other. Two north poles are the same, or **like poles**. *Like poles* push each other away.

Like poles: The same poles; two north or two south poles.

In which pictures will the poles pull on each other? In which pictures will they push away from each other?

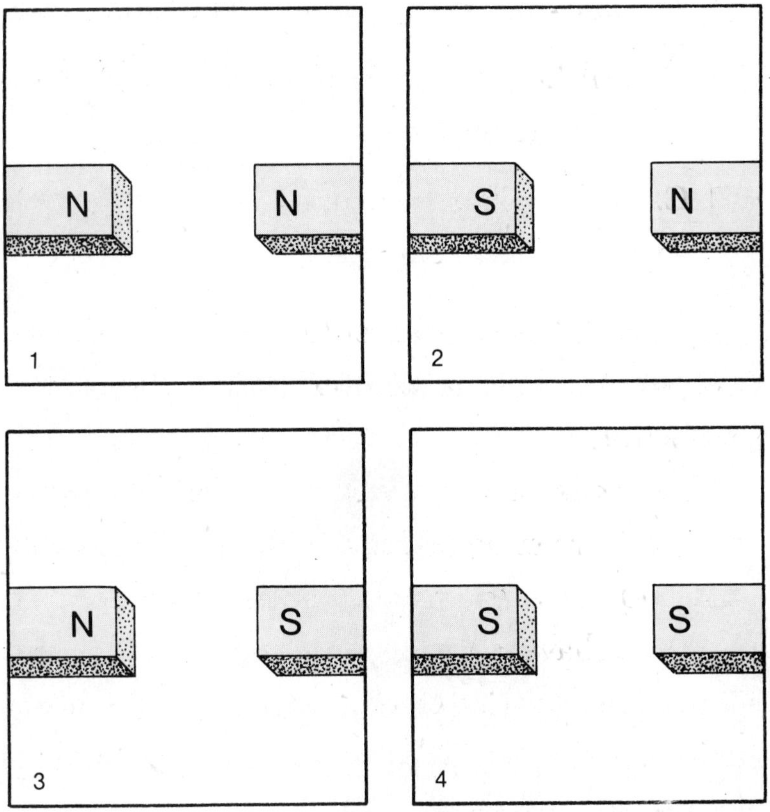

ACTIVITY

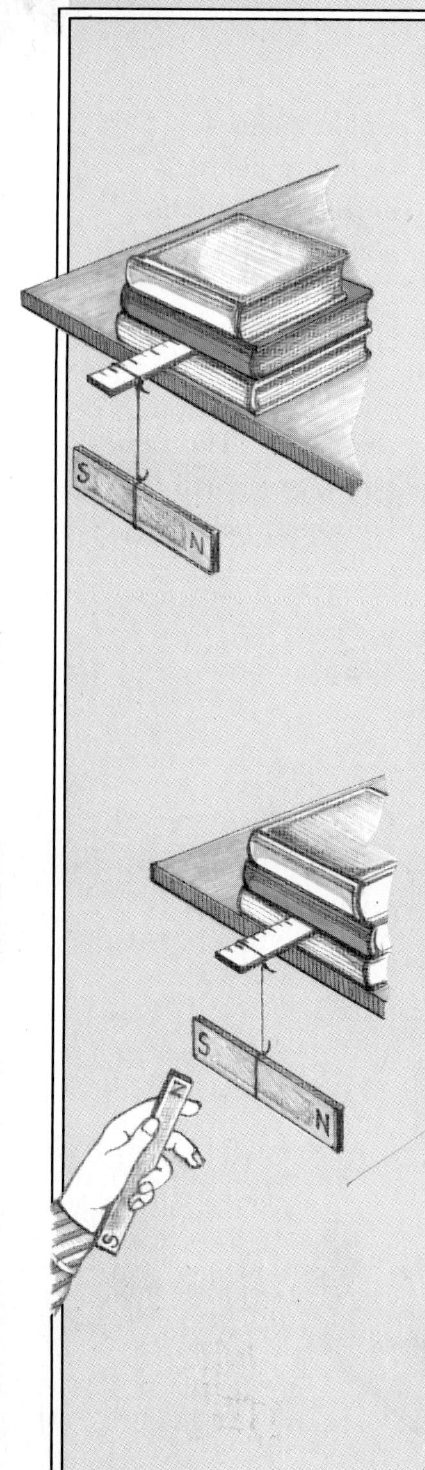

How do the poles of magnets act?

A. Gather these materials: 2 bar magnets (labeled N and S), 3 books, ruler, and string 30 centimeters long.

B. Tie one end of the string to the end of the ruler. Tie the other end to the middle of 1 bar magnet.

C. Place the books at the edge of your desk.

D. As shown in the picture, place the ruler between the books. Make sure the hanging magnet is not moving.

E. Bring the north pole of the other bar magnet near the south pole of the hanging magnet.

1. What did you see?

F. Bring the south pole of the magnet near the north pole of the hanging magnet.

2. What did you see?

G. Bring the north pole of the magnet near the north pole of the hanging magnet.

3. What did you see?

4. What do you think will happen if you repeat step G, using the south pole?

H. Bring the south pole of the magnet near the south pole of the hanging magnet.

5. What did you see?

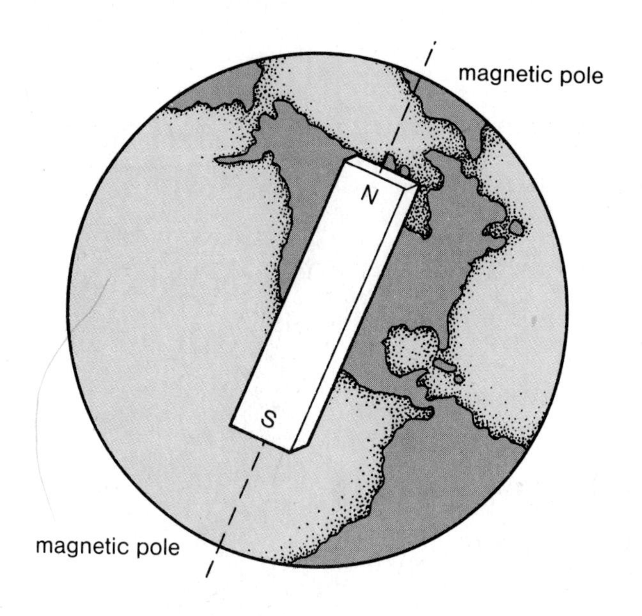

The earth is like a huge magnet. It has a north and a south magnetic pole. Long ago, people found a way to find north and south on earth. They used stones with a lot of iron in them. The stones acted like magnets. They pointed north.

Today, people use a **compass** (**kum**-pus) to tell direction. Have you ever used a *compass* like the one in the picture below?

Compass: An object with a magnet that helps you tell direction.

A compass has a metal needle. The needle is a magnet. A compass has the letters N, S, E, and W printed on the face. N stands for north. S stands for south. Do you know what E and W stand for? North, south, east, and west are the directions on a compass. A compass needle always points north.

Section Review

Main Ideas: The ends of a magnet are called poles. A magnet has a north pole and a south pole. A magnet is strongest at its poles. Unlike poles pull on each other. Like poles push each other away. The north-seeking pole of a compass needle points to the north magnetic pole of the earth.

Questions: Answer in complete sentences.

1. What are the ends of a magnet called?
2. What part of a magnet is the strongest?
3. What happens when unlike poles are brought near each other?
4. What happens when like poles are brought near each other?
5. Explain how the magnet of a compass tells direction.
6. Look at the pictures on page 197. What will happen in each picture if the magnets are brought together?

CHAPTER REVIEW

Science Words: The clues in column B will help you unscramble the words in column A.

Column A	Column B
1. SSAPCOM	An object with a magnet that helps you tell direction
2. SOLEP	The ends of a magnet
3. NIOR	A metal that can be picked up by a magnet
4. EELTS	A metal made from iron that can be picked up by a magnet
5. STENGAM	Objects that pick up or stick to certain materials

Questions: Answer in complete sentences.

1. If a magnet picks up an object, what is the object made of?
2. What is a compass? What is it used for?
3. Look at this picture. Where is the magnet the strongest? What must the paper clips be made of?

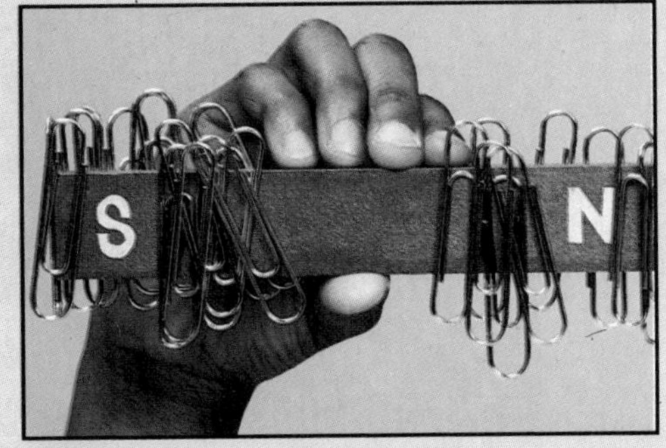

4. What do the letters N, S, E, and W stand for on a compass?
5. In which direction does a compass needle point?
6. Draw a picture of a bar magnet. Label the ends.

CHAPTER 14

ELECTRICITY

14-1. Static Electricity

Have you ever seen streaks of light like these during a storm? These streaks of light in the sky are called lightning. Do you know what causes lightning?

When you finish this section, you should be able to:

- ☐ **A.** Show how *electric charges* can move very light objects.
- ☐ **B.** Give examples of *static electricity.*

All matter contains tiny units of electricity. These units are **electric charges** (**charj**-ez). *Electric charges* can be rubbed off one object and put onto another object. Have you ever rubbed a plastic ruler with wool? What happened when you brought the ruler near small pieces of paper? The charges on the ruler pulled the paper toward the ruler.

Electric charges: Tiny units of electricity.

When an object gains or loses charges, it has **static** (**stat**-ik) **electricity.** Lightning is caused by *static electricity.* When electrical charges jump from one cloud to another or from a cloud to the earth, there is a flash of light.

Static electricity: Electricity caused when objects lose or gain charges.

A plastic or hard-rubber comb can pull charges from your hair. When this happens, each hair becomes charged with electricity. The hairs push each other apart. This makes it hard to keep your hair in place.

ACTIVITY

Can you see static electricity?

A. Gather these materials: paper, pencil shavings, plastic ruler, clear tape, and 15-centimeter wool square.

B. Place the pencil shavings on a sheet of paper. Rub the wool back and forth against the ruler. Do this about 10 times. Hold the ruler over the pencil shavings.

1. What did you see? Record your findings on a sheet of paper.

C. Cut 2 pieces of tape, each 15 centimeters long. Stick the pieces onto your desk. Leave a small part hanging over the edge. You will soon pull the tape off the desk. Then, you will hold the 2 pieces near each other.

2. What do you think will happen?

D. Now, hold the ends of the tape. Pull the tape off the desk. Bring the pieces near each other.

3. What happened?

4. Did you think this would happen? Explain.

Section Review

Main Ideas: All matter contains electric charges. Charges can be rubbed off objects. When an object gains or loses charges, it has static electricity.

Questions: Answer in complete sentences.

1. What type of electricity is caused by an object losing or gaining charges?
2. What happens when you rub wool against a plastic comb?

3. The small pieces of paper are sticking to the balloon in the above picture. Answer these questions about the picture.
   a. What do you think the girl did to the balloon before holding it near the papers?
   b. Did the balloon gain or lose charges?
   c. What caused the papers to stick to the balloon?

14-2. Electric Current

Mark and Sarah are getting ready to go camping. They are deciding which flashlight to take. They need one that will give out a lot of light. It must also last a long time. Which one would you choose?

When you finish this section, you should be able to:

- ☐ **A.** Tell what happens when charges move.
- ☐ **B.** Describe what is needed for charges to move from place to place.
- ☐ **C.** Name some materials that carry charges and some that do not.

Mark and Sarah need a flashlight with many **dry cells**. A *dry cell* is a source of electric charges. It makes the electricity in a flashlight. A dry cell is sometimes called a battery (**ba**-ter-ee). Charges from the dry cell travel through metal to the bulb. The charges light up the bulb.

Dry cell: A source of electric charges.

Moving electric charges are called **electric current** (ih-**lek**-trik **ker**-ent). *Electric current* is made when charges move from place to place.

Electric current: Moving electric charges.

Look at the picture below. It shows the path charges follow from a power station to your classroom.

When a current moves from a source to a user, a **circuit** (**ser**-kit) is made. A *circuit* is the path that lets charges move from place to place.

Circuit: A path for moving charges.

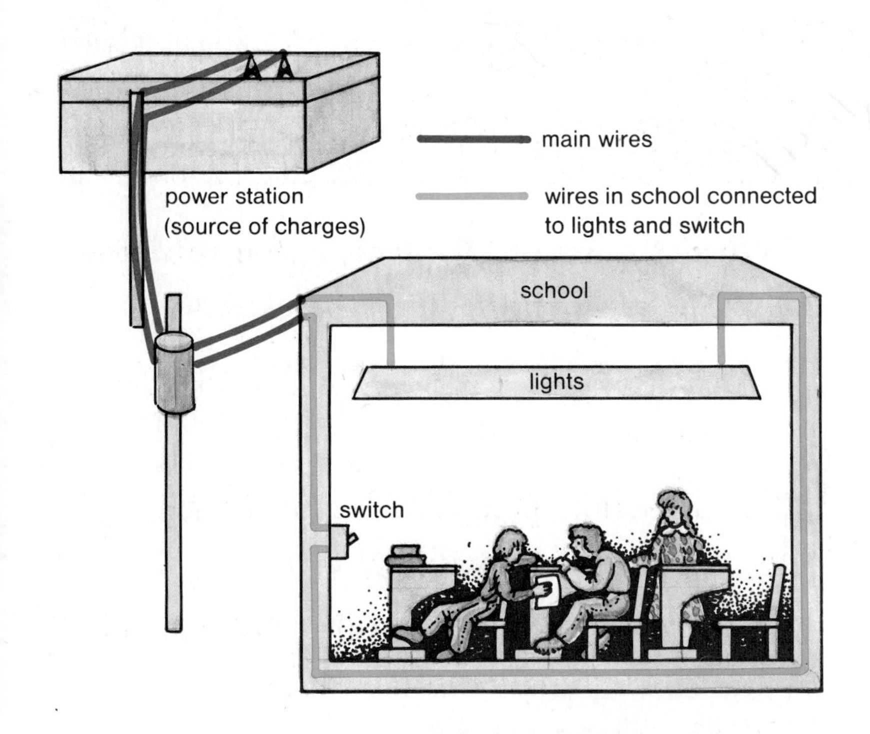

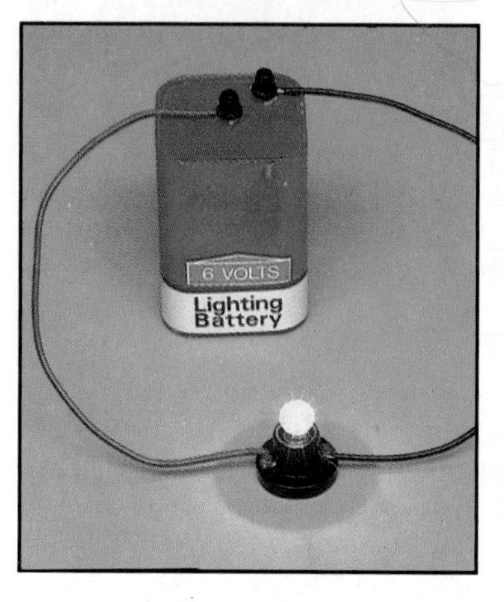

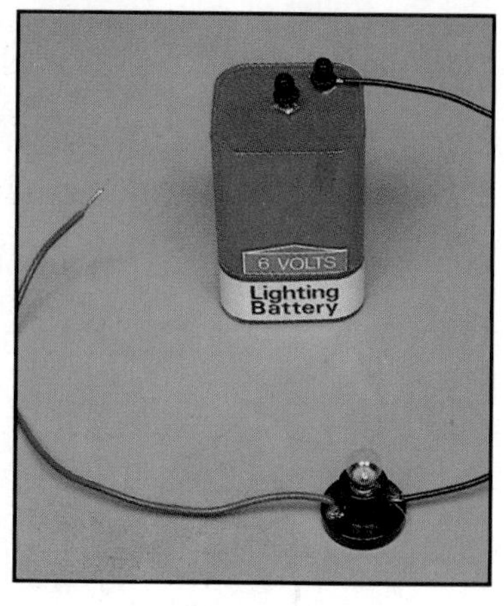

When a switch is turned on, current can get to a light bulb. This happens because all parts of the circuit are then connected. When all parts are connected, the circuit is called a **closed circuit**. Current can move only through a *closed circuit*. When the switch is turned off, the light goes off. With the switch off, parts of the circuit are not connected. The circuit is then called an **open circuit**. Current cannot move through an *open circuit*. Which picture shows an open circuit?

Closed circuit: A circuit in which all parts are connected.

Open circuit: A circuit in which not all parts are connected.

Electric current can move through some materials easily. These materials are called **conductors** (kun-**duk**-terz). Most metals are good *conductors*. Copper and iron are metals. Your body is also a good conductor.

Conductors: Materials that let current move through them easily.

Look at the picture at the left on page 209. Is the circuit open or closed? The wires are touching the key. The current moves through the metal key as if it were wire. The key is a good conductor. Look at the next picture. Is the circuit open or closed?

ACTIVITY

Can you make a circuit?

A. Gather these materials: D-size battery, a bulb and socket, 3 pieces of wire (30 centimeters long), key, toothpick, and clear tape.

B. Tape one bare end of the wire to the bottom of the battery. Be sure the wire is taped tightly, so it will not fall off.

C. Wrap the other bare end of the wire tightly around the metal part of the bulb.

D. Touch the end of the bulb to the top of the battery. Then remove it.

1. What did you see? Record your findings.

2. When was the circuit open? When was it closed?

E. Look at the picture below. It shows how to find out if things can conduct electricity. Test the key and the toothpick.

3. Was the key or toothpick a conductor?

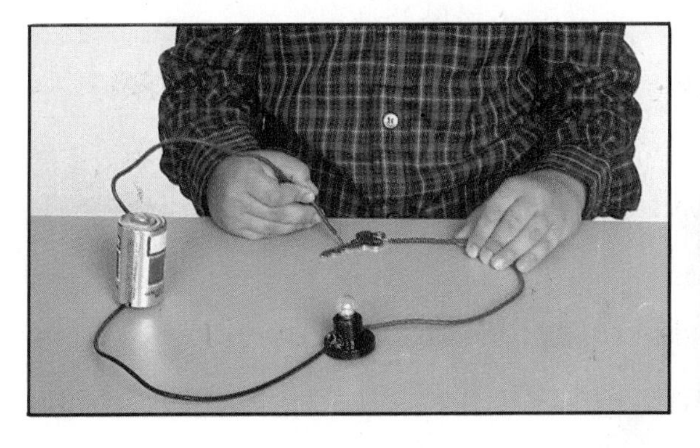

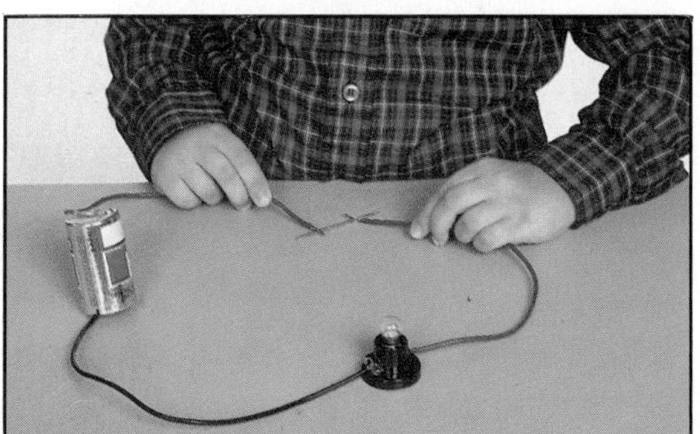

Insulators: Materials that do not allow current to move through them easily.

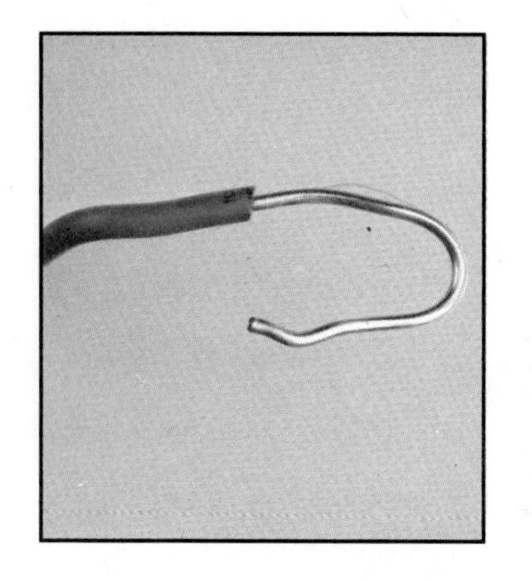

Materials that do not allow current to move through them easily are **insulators** (**in**-suh-lay-terz). Paper is an *insulator*. Can you name some other materials that are insulators? Glass, rubber, cloth, wood, and plastic are other insulators. Sometimes, conductors are covered with insulators. Do you know why most electric wires are covered with rubber or plastic? These materials keep you from getting a shock when current flows through the wires. Insulators also stop charges from leaving a conductor. Look at the picture at the left. The insulator is rubber. What kind of metal is the conductor?

Section Review

Main Ideas: A circuit is a path that lets electric charges move from place to place. Electric current can move easily through conductors. A current cannot move easily through insulators.

Questions: Answer in complete sentences.

1. What are moving electric charges called?
2. Name the parts of a circuit.
3. What will happen to current if a circuit is opened?
4. What will happen to current if a circuit is closed?
5. How do a conductor and an insulator differ? Name a conductor and an insulator.

14-3. Safety with Electricity

Electricity is very helpful. It lights our homes and schools. It helps us cook food, freeze food, and heat buildings. It helps us do many other things. But electricity can be very dangerous.

When you finish this section, you should be able to:

☐ **A.** Explain how to prevent fires caused by electricity.

☐ **B.** List rules to follow when using electricity.

Electricity can help you do many things. It is also dangerous. It can kill you in a second. If you touch a bare, current-carrying wire, charges will jump from the wire to your body. You will get an electric shock.

Does an outlet in your home or school look like the one in the picture above? If so, it may cause a fire.

Never plug too many things into an outlet at one time. Too many charges will move through the wires. The wires can get very hot. They may get hot enough to start a fire.

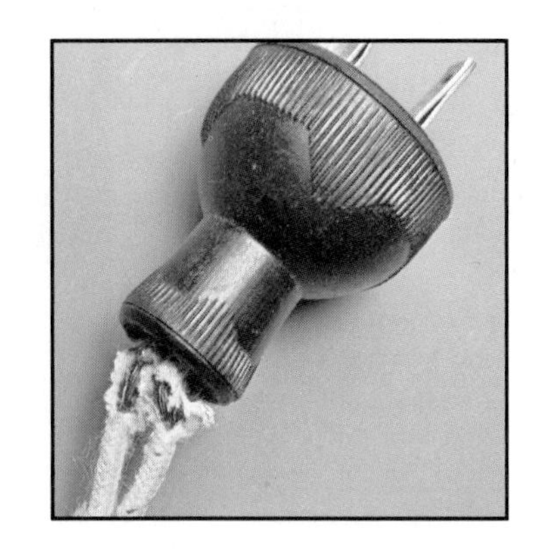

Look at the wires in this picture. Part of the insulator is worn off. Someone might touch this bare wire. If you see plugs or wires like this, do *not* touch them. Tell an adult. The adult will remove them without touching the bare wire.

Here are some rules to follow for the safe use of electricity:

1. Never touch anything that uses electricity when in the bath or when hands or feet are wet.

2. Never stick your finger or an object into an outlet.

3. Turn off lights and other things that use electricity when they are not in use.

ACTIVITY

Can you make picture posters on electrical safety?

A. Gather these materials: 3 sheets of poster paper, watercolor marking pens or paint, metric ruler, and pencil.

B. Pick 3 safety rules. Make a safety poster for each rule. Write out the rules. Then, draw cartoons showing what happens if you do not follow the rules. You can add short sayings to help you remember each rule. For example: "Don't switch on fans with wet hands."

C. List places where the posters could be used.

4. Do not try to fix anything that uses electricity. Let an adult do it.

5. Never fly a kite near electric wires.

6. Never use a metal ladder near electric wires. The ladder may touch the wires.

7. Do not put electric wires under rugs or through doorways. The insulation can wear off. A spark can jump from the bare wire. Heat from a wire under a rug also might start a fire.

8. Stay away from places with signs that say "High Voltage."

9. Do not stand under or near a tree or pole during an electric storm. Lightning might strike the tree or pole and jump onto your body.

10. Stay away from electric wires that have fallen to the ground.

11. Never touch the battery in a car.

Section Review

Main Ideas: Electric current can be very useful to us. But it can also be dangerous. People should always follow safety rules when using electricity.

Questions: Answer in complete sentences.

1. A picture in this section shows a safety problem caused by a wire. What page is this picture on?
2. Look at the picture you found. Which safety rule should a person follow?
3. Which safety rules should the child in each of these pictures follow?

CHAPTER REVIEW

Science Words: Write the sentences below on paper. Fill in the blanks with the correct words from the list.

dry	**circuit**	**closed**
open	**charges**	

1. Tiny electric ______ are found in all matter.
2. A ______ is a path for moving charges.
3. When the parts of a circuit are connected, the circuit is ______.
4. When the parts of a circuit are *not* connected, the circuit is ______.
5. A ______ cell is a source of electric charges.

Questions: Answer in complete sentences.

1. How is a conductor different from an insulator?
2. What are three things a circuit must have?
3. Name a conductor.
4. Name an insulator.
5. List three safety rules for using electricity.
6. Which of the circuits below is an open circuit?
7. Which of the circuits below is a closed circuit?

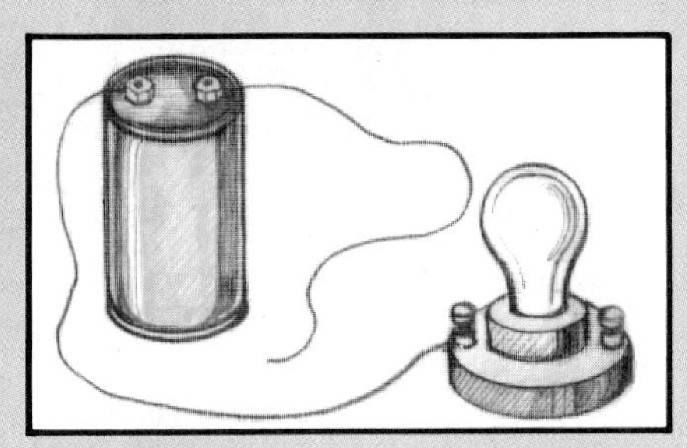

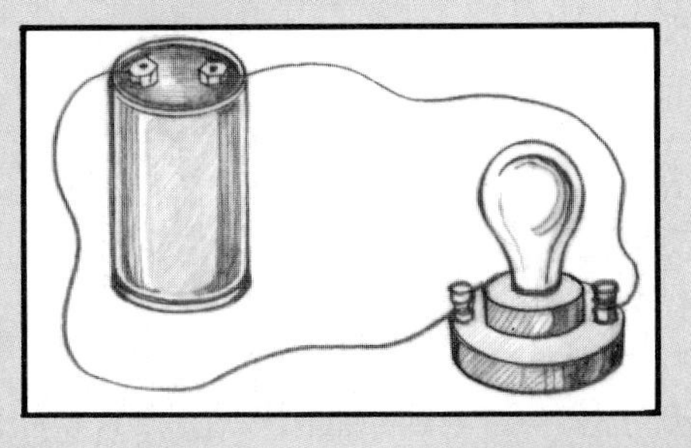

CHAPTER 15

USING ELECTRICITY

15-1. Changing Electricity

We use electricity in many ways. Look at the picture above. How is each child using electricity? Each child is using something that changes electricity.

When you finish this section, you should be able to:

☐ **A.** Explain how electricity can be changed to other types of energy.

☐ **B.** Give an example of something that changes electricity.

When we use electricity, it must be changed in some way. A light bulb changes electricity into light. Do you see the thin wire inside this bulb? The thin wire is called a **filament** (**fil**-uh-ment). Electricity moves through the *filament*. The filament gets hot and glows. It gives off light.

Filament: A thin wire in a light bulb.

Electricity moves through the wires in a toaster. This heats the wires. The wires give off heat and light. The heat toasts the bread.

Electromagnet: A magnet made by electricity.

Do you know why the blades of a fan and the hands on an electric clock move? They move because electricity can be used to make a magnet. Such a magnet is an **electromagnet** (ih-lek-troh-**mag**-net).

An *electromagnet* is a piece of iron wrapped with wire. When a current passes through the wire, the iron becomes a magnet. When the current is shut off, the magnetism stops. Electric motors use electromagnets. Electric trains, fans, and clocks all have electric motors.

An electric doorbell changes electricity into sound. When you press the button, a sound is made. When you press the button, current flows. The current goes through the wires around a bar of iron. The iron becomes an electromagnet. The electromagnet pulls a metal piece against the bell. Can you find the electromagnet in this picture?

ACTIVITY

Can you build a magnet that uses electricity?

A. Gather these materials: 1 1/2-volt (1.5-volt) dry cell, insulated bell wire (45 centimeters long), compass, 6 paper clips, clear tape, and steel nail.

B. Your teacher will remove about 2 1/2 centimeters of insulation from each end of the wire.

C. Tape one end of the bell wire to the bottom of the dry cell.

D. Wrap the wire around the nail about 15 times.

1. What do you think will happen if you bring the compass near the nail?

E. Move the compass in a circle near one end of the nail.

2. What did you see?

F. Move the compass away. Touch the free end of the wire to the top of the dry cell. Now, bring the compass near the dry cell.

3. What did you see?

4. Do you think your electromagnet can pick up the paper clips?

G. Try to pick up the paper clips.

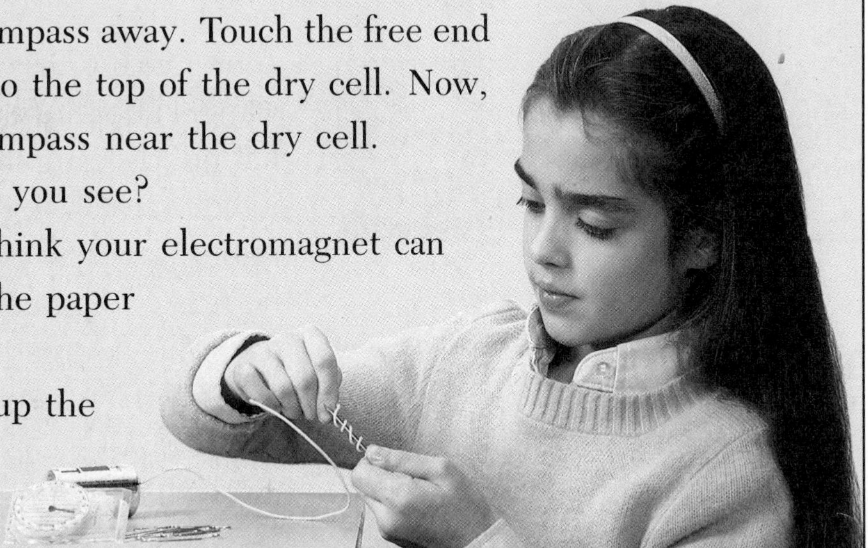

Section Review

Main Ideas: Electricity can be changed into heat, light, sound, and magnetism. Electric motors use electromagnets.

Questions: Answer in complete sentences.

1. Explain how electricity is changed into light by a light bulb.
2. Explain how electricity is changed into sound by a doorbell.
3. Name something that changes electricity into energy.

An Wang

An Wang was born in Shanghai, China. He came to this country in 1945. He studied physics at Harvard University. In 1948, Dr. Wang invented a part used in computers. The part is called a magnetic core. It is the part of the computer that stores information.

In 1951, Dr. Wang started Wang Laboratories, a company that builds computers. The computers are used in schools, homes, and offices. They were the first computers to use a television-like screen. This screen shows what has been typed into the machine. Dr. Wang's company also builds computers in China.

15-2. Making and Conserving Electricity

What are these children looking at? You may have seen something like it at home or at school. Do you know what it does?

When you finish this section, you should be able to:

- ☐ **A.** List ways we can save electricity.
- ☐ **B.** Describe how electricity is made.
- ☐ **C.** Describe three ways of producing electricity without using up fuel.

The picture above shows an electric meter. It measures how much energy is used in a building. Someone from the electric company reads the meter each month. The electric company sends a bill to the people in the building. Electricity costs money. If people use a lot of electricity, they get big bills.

We should not waste electricity. That is why we should turn off radios, televisions, and lights when no one is using them.

Fuel: Something that can be burned.

Electricity is made mostly by burning **fuel**. *Fuel* is something that can be burned. Coal, gas, and oil are fuels. When fuel burns, it produces heat. The heat runs machinery that makes electricity.

It took millions of years for fuels to form. Fuels were made when plants were covered by rock long ago. Fuels are found deep underground. It costs money to find and get fuels. People who take fuel from the ground have hard jobs. These jobs can also be dangerous.

Coal, oil, and gas can be saved if we find new ways to make electricity. One way is to use the power of the wind. A second way is to use sunlight. What is used to make electricity in each of the above pictures?

A material called **uranium** (yoo-**ray**-nee-um) is used in some places. It does not burn. *Uranium* is changed into heat energy in a **nuclear reactor** (**noo**-klee-er ree-**ak**-ter).

Uranium: A material that does not burn but can be used to make electricity.

A little uranium can make a lot of heat. The heat is then used to make electricity. *Nuclear reactors* also make energy that can harm people, animals, and plants. For this reason, some people believe nuclear reactors should not be used. Other people think we need them because we may run out of other fuel.

Nuclear reactor: A place where electricity is made by using uranium.

ACTIVITY

How can electricity be saved in your school or home?

A. Gather these materials: paper, pencil, and metric ruler.

B. Make a chart showing things that use electricity in your school or home. Your chart should include: (1) the name of the object, (2) a drawing of the object, (3) what the object does, (4) who uses the object, (5) one idea about how the object could be used less often.

Section Review

Main Ideas: Electricity is made from fuel. When we save electricity, we save fuel. People are trying to find new ways to make electricity.

Questions: Answer in complete sentences.

1. Name two materials used to make electricity.
2. List two ways to make electricity without using fuel.
3. What is one way to save electricity?
4. A picture in this section shows people working hard to get fuel. Find the picture. Then, write a sentence about it.

These children are using a machine to help them learn. Do you know what it is called?

15-3.

Computers, Bionic Arms, and Other Amazing Things

When you finish this section, you should be able to:

- ☐ **A.** Tell how a *computer* helps people solve problems.
- ☐ **B.** Describe four parts of a *computer*.
- ☐ **C.** Explain how *bionic* inventions help sick and injured people.

The children in the picture are using a **computer** (kum-**pyoo**-ter). A *computer* uses electricity to solve problems. The computer in the picture gets electricity from an outlet.

Computer: An invention that helps people solve problems.

Activity

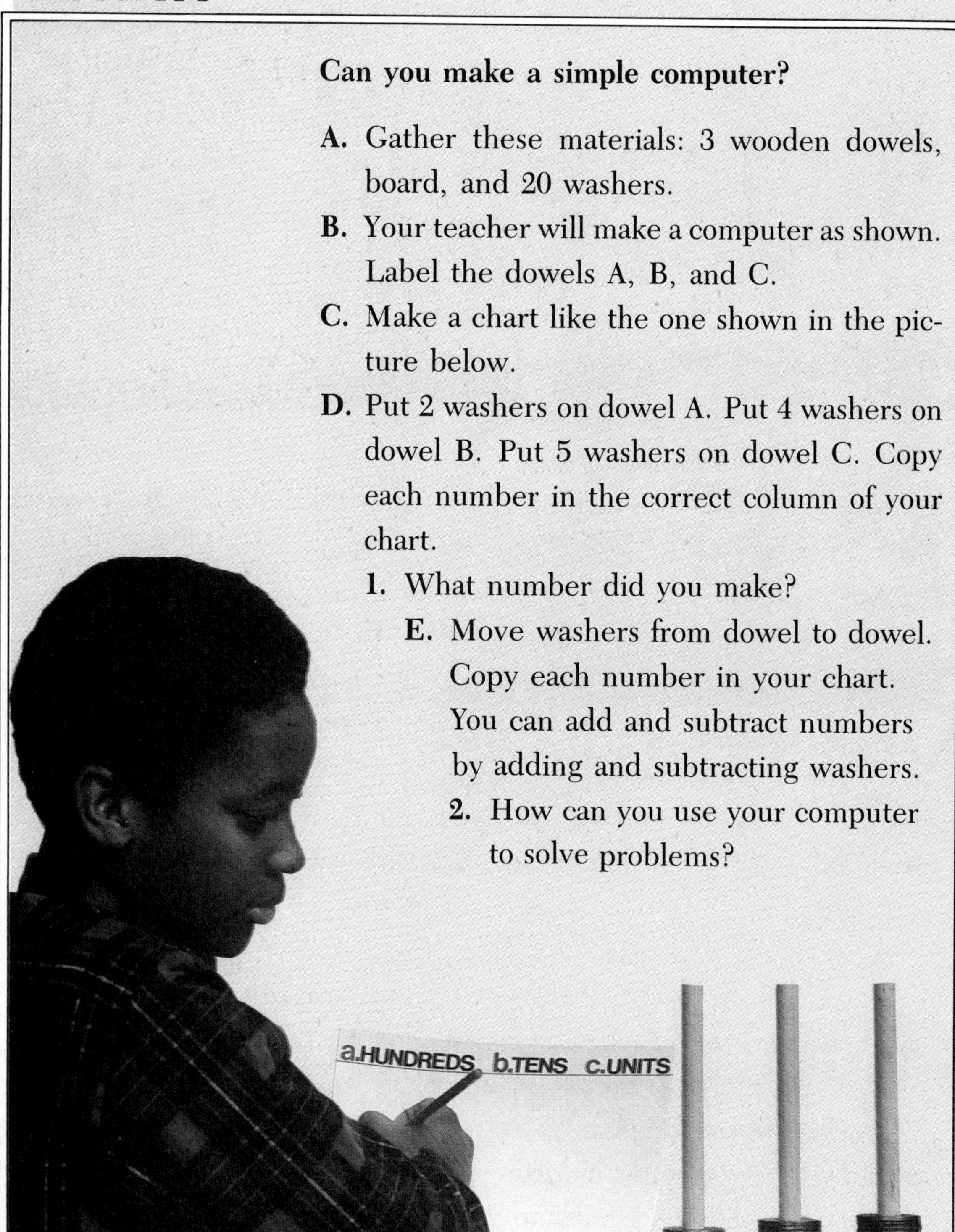

Can you make a simple computer?

A. Gather these materials: 3 wooden dowels, board, and 20 washers.

B. Your teacher will make a computer as shown. Label the dowels A, B, and C.

C. Make a chart like the one shown in the picture below.

D. Put 2 washers on dowel A. Put 4 washers on dowel B. Put 5 washers on dowel C. Copy each number in the correct column of your chart.

1. What number did you make?

E. Move washers from dowel to dowel. Copy each number in your chart. You can add and subtract numbers by adding and subtracting washers.

2. How can you use your computer to solve problems?

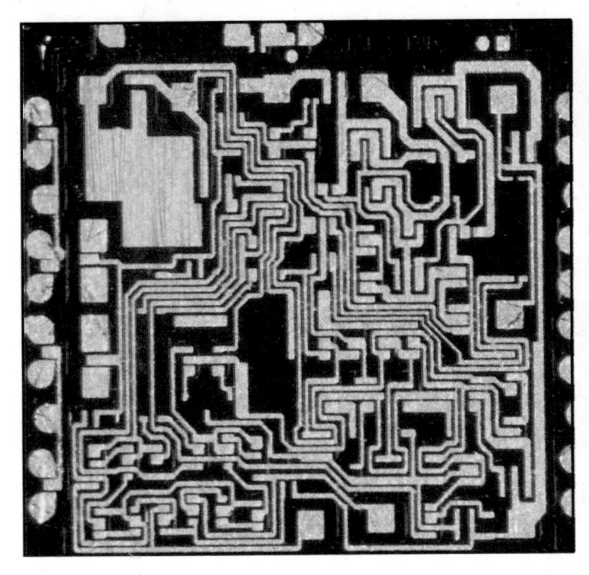

One part of the computer is its **keyboard**. The *keyboard* is the place where you type information, such as words or numbers.

Keyboard: The part of a computer that you type on.

The keyboard sends information through wires into **chips**. A *chip* is a tiny part in a computer. It has many electric circuits. The picture on the left above shows a computer chip. The picture has been enlarged, so you can see the parts of a chip. The picture on the right has also been made larger. It shows a chip on a penny.

Chip: A tiny part of a computer; it has many circuits.

The computer part that shows what you have typed is the **video** (**vih**-dee-oh) **screen**. The *video screen* also shows a computer's answers.

Video screen: The part of a computer that shows you what you typed; it also shows a computer's answers.

Sometimes you need a copy of what is on the video screen. A **printer** makes a copy for you. A *printer* uses electricity to make marks on a sheet of paper.

Printer: A computer part that makes copies of what you typed or of the computer's answers.

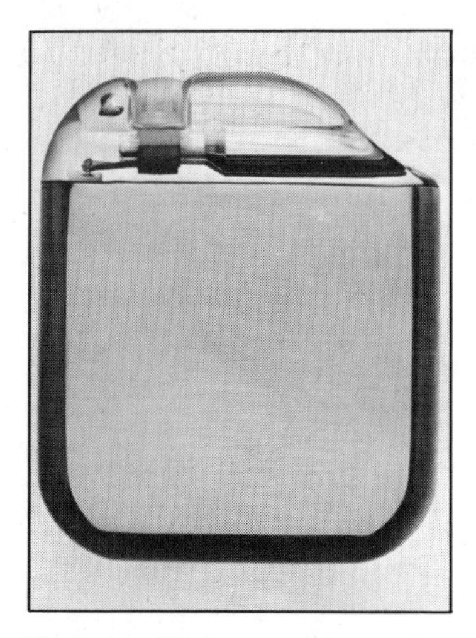

Bionic: Using electricity to help the body work.

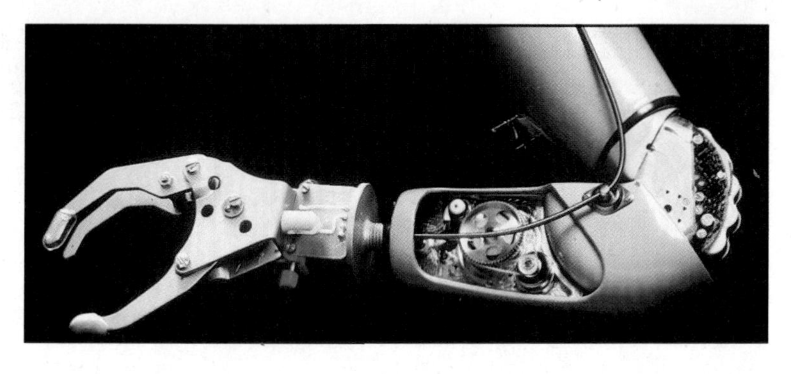

A computer is only one invention that uses electricity. Another one is shown above at the right. This is a **bionic** (by-**ahn**-ik) arm. A *bionic* arm uses electricity. It replaces a real arm. Tiny dry cells, chips, and motors make the bionic arm work. This arm is being tested now.

The picture at the left shows another bionic invention. It helps people with heart problems. The pacemaker is connected to the heart. It sends out tiny amounts of electricity every few seconds. This electricity makes the heart beat.

Section Review

Main Ideas: Electricity is used in many inventions. Computers and bionic arms are examples.

Questions: Answer in complete sentences.

1. Name and describe four parts of a computer.
2. What machine described in this section helps people solve problems?
3. Name two bionic inventions. Tell how they work.

CHAPTER REVIEW

Science Words: Unscramble the letters to find the correct terms. Then write what each term means.

1. seluf
2. tanemilf
3. teelectrongam
4. icboin
5. anmuiur

Questions: Answer in complete sentences.

1. Why should we try to use less electricity?
2. What would you need to make an electromagnet?
3. What kinds of objects can electromagnets pick up?
4. A light bulb changes electricity into light and heat. Explain how a light bulb does this.
5. Name something that changes electricity into sound.
6. What is the thin wire inside a light bulb called?
7. Name the part of a computer that shows its answers.
8. Name three fuels.
9. What is a chip?
10. How does a toaster work?
11. Why do some people think that nuclear reactors should not be used to make energy?
12. What part of a computer makes copies of what you have typed into it?

INVESTIGATING

Making a circuit mystery box

A. Gather these materials: shoe box, 9 paper fasteners, 10 lengths of wire (each 10 centimeters long), flashlight bulb, flashlight-bulb holder, dry cell, 4 lengths of wire (each 15 centimeters long), and tape.

B. Make a circuit tester like the one shown. Place the fasteners through the shoe box top. Use a few lengths of 10-centimeter wire to attach some of them to the underside of the box top.

C. Put the top back on the box. Touch the free ends of the wire to the tops of 2 fasteners.

1. What do you see?
2. How can you tell which fasteners are connected without opening the box?

D. Add other wires to connect more fasteners. Exchange boxes with another student. Use your circuit tester to predict which fasteners are connected.

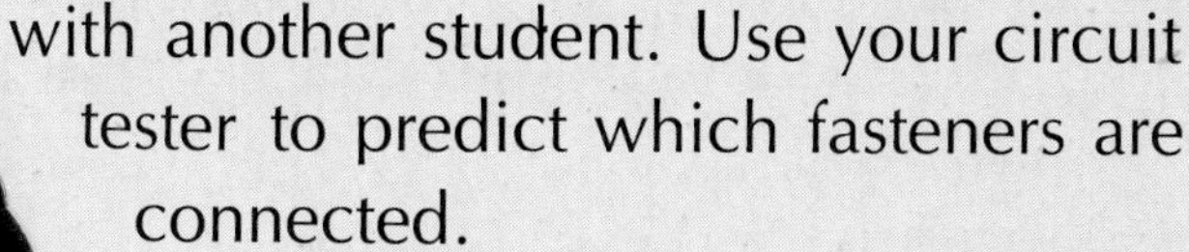

3. Make a drawing showing how the fasteners are connected.
4. Open the box. Draw the pattern the wires form.

CAREERS

Computer Programmer ▶

You have been learning about computers. A computer programmer writes step-by-step orders that tell the computer how to solve problems. Programmers also write orders to make the computer change answers into forms that can be understood. Computer programmers need technical training.

◀ Electronic Technician

You have been learning about electronic machines. Electronic technicians test new electronic products. They also write books that tell how these products work. Some technicians repair these machines. Special training is needed for a job as an electronic technician.

WHERE PLANTS AND ANIMALS LIVE

UNIT 6

CHAPTER 16

THE FOREST AND THE GRASSLAND

16-1. The Forest

Have you ever stood under a big tree in the summer? If you have, then you know that it is shady and cool there. A place with many trees is a forest. Do you know what else lives in a forest?

When you finish this section, you should be able to:

- ☐ **A.** Name plants that live in a forest.
- ☐ **B.** Name animals that live in a forest.

Do you know what a community (kuh-**myoo**-nih-tee) is? A community is a group of plants and animals living in one place.

A forest is a type of community. In a forest, a group of plants and animals live together. The largest plants in a forest are trees. The leaves of the tall trees in a forest are like a huge umbrella. Many other plants live in the shade of this leafy umbrella. Grass does not grow on a forest floor. Instead there are mosses and other plants that can grow in shade. Moss is a small green plant that grows in damp soil.

Plants live where they can find the right soil and the right amount of rainfall. The temperature and the amount of sunlight must also be right. Some forests are evergreen forests, with trees like pines. Other forests are made up mainly of trees that lose their leaves each year. Why do you think there are different kinds of forests? The differences in temperature are one reason. The amount of rain and light is another reason.

Shelter: A place that protects or covers an animal.

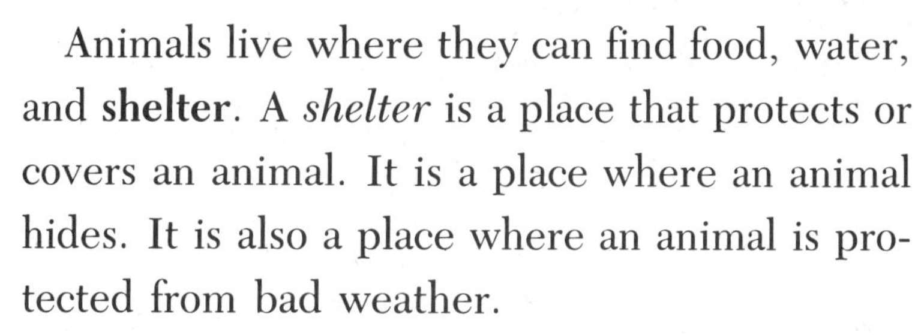

Animals live where they can find food, water, and **shelter**. A *shelter* is a place that protects or covers an animal. It is a place where an animal hides. It is also a place where an animal is protected from bad weather.

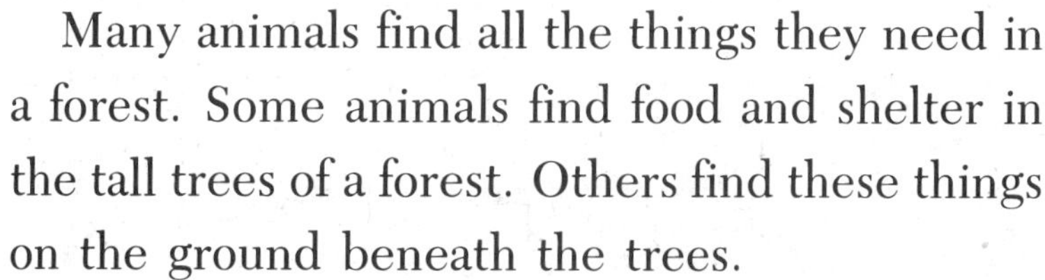

Many animals find all the things they need in a forest. Some animals find food and shelter in the tall trees of a forest. Others find these things on the ground beneath the trees.

Do you recognize the small bird shown above at the left? It is a chickadee (**chih**-kuh-dee). Chickadees live in the forest. They build nests in hollow trees or branches. They use moss to build their nests. Chickadees eat insects, insect eggs, and seeds. One food they really like is bark beetles. These beetles live in the bark of trees.

An owl is a large bird that lives in the forest. An owl has very large eyes. At night, the owl hunts for small forest animals, such as mice and birds.

What do you think you would find under a branch that has fallen on the forest floor? Perhaps it would be a salamander (**sal**-uh-man-der). A salamander looks like a lizard. It has soft skin and no claws. Some salamanders live in water. Others live on land. Salamanders eat insects, worms, and other small animals that live in the soil.

Deer are among the largest animals that live in forests. In the winter, deer eat mosses, buds, nuts, and berries. During other seasons, they eat twigs, grass, and leaves. Deer like to rest in leafy areas. Do you know why? One reason is that these are good hiding places.

Burrow: A tunnel under the ground.

A noisy forest animal is the chipmunk. A chipmunk eats insects, seeds, nuts, and berries. It looks for these things under leaves on the forest floor. When a chipmunk is alarmed, it chatters very loudly. Then it disappears. Do you know where it goes? Sometimes it hops onto a tree. Or it might go into its **burrow** (**buh**-roh). A *burrow* is a tunnel under the ground. A chipmunk's burrow has many tunnels. Look at the picture at the top of page 239. Where does a chipmunk enter its burrow? Where does it sleep? Where does a chipmunk keep its food? What would happen if another animal went into the burrow? Where would the chipmunk go? Do you think a burrow is a good shelter for a chipmunk? In the next section, you will read about another animal that lives in a burrow.

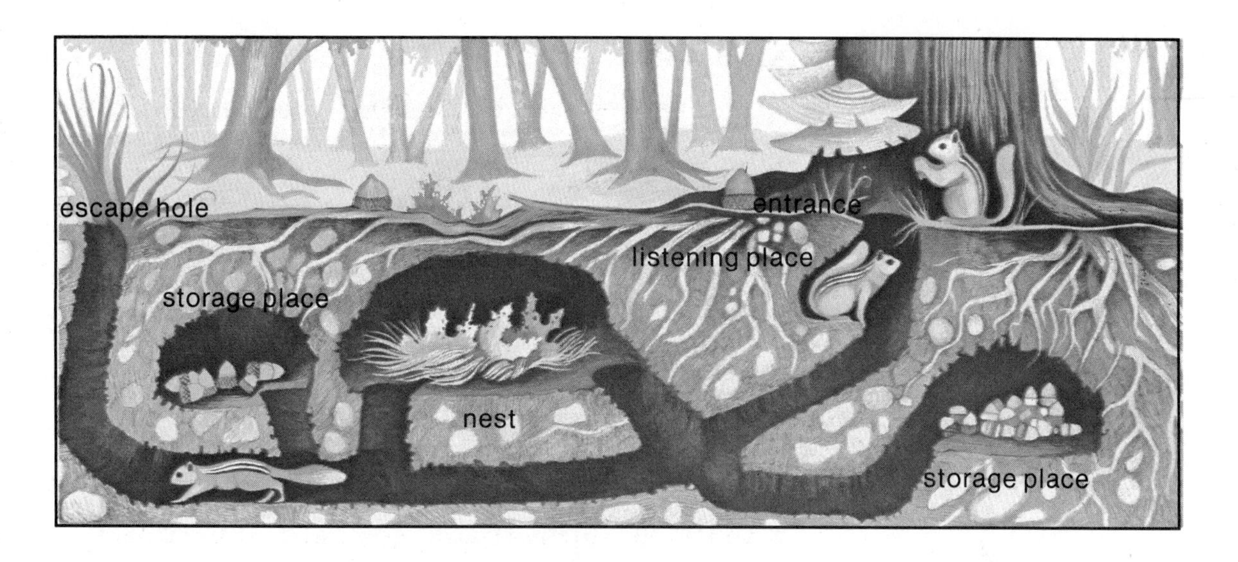

A chipmunk lives in the forest because it can find food, water, and shelter there. The forest is a chipmunk's **habitat** (**hab**-uh-tat). A *habitat* is a

Habitat: A place where a plant or animal lives.

ACTIVITY

Can you make a habitat for small animals?

A. Gather these materials: large jar, soil, bits of leaves and roots, water, and earthworms.
B. Fill the jar two-thirds full of soil.
C. Water the soil.
D. Sprinkle the leaves and roots on the soil.
E. Put the worms in the jar.
  1. What do the worms do?
F. Put the jar in a cool, dark place. Look at it each day.
  2. What is happening inside the jar?
  3. What do the worms eat?

place where a plant or animal lives. A community of plants and animals can share the same habitat.

Section Review

Main Ideas: Plants live where the soil, temperature, sunlight, and amount of rainfall are right for them. Animals live where they can find food, water, and shelter. A forest is a community of many plants and animals.

Questions: Answer in complete sentences.

1. What is a forest?
2. What four things do plants need to live?
3. What three things do animals need to live?
4. Name three animals that live in a forest.
5. Why is it important for animals to have a shelter?

16-2. The Grassland

Long ago there were many bison (**by**-sun) in large areas of our country. American bison are larger than cattle. They are sometimes called buffalo (**buh**-fuh-loh). The areas where the bison lived had very few trees. These areas were covered with grass. The bison fed on the grass. Today, there are farms and ranches in these grassy areas. These areas are called grasslands. What is a grassland like? Do you know what animals live in the grasslands today?

When you finish this section, you should be able to:

- ☐ **A.** Describe a grassland.
- ☐ **B.** Name some plants that live in a grassland.
- ☐ **C.** Name some animals that live in a grassland.

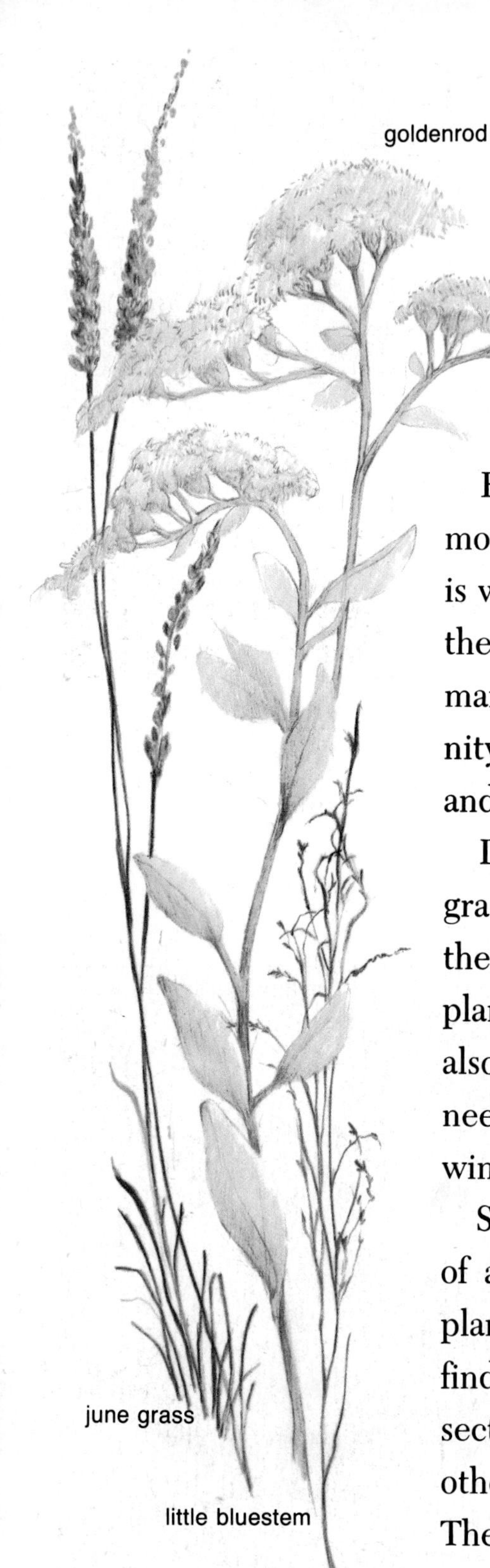

From a distance, a grassland may look like a moving sea of grass. Everywhere you look, grass is waving in the wind. All the grass seems to be the same. But if you look closely, you will find many kinds of plants. A grassland is a community of different kinds of grasses, wildflowers, and other plants.

Do you know why there are not many trees in grassland areas? There is not enough rain. But there is enough rain for grass and many other plants. Grasslands are sunny places. They can also be very windy. The plants of a grassland need a lot of sun. They are not harmed by the wind.

Suppose you built a fence around a small part of a grassland. You would find many kinds of plants. What else would you find? You might find thousands of grasshoppers and other insects. There would probably also be birds, mice, other small animals, and a few large animals. These animals can find food and shelter in a grassland.

A grasshopper eats grass. With its strong mouth, a grasshopper can easily cut a lot of grass in one day. It also eats other plants. Look at the picture of a grasshopper. How does its color protect it from other animals?

Another animal that lives in the grassland is the harvest mouse. This small mouse eats seeds. A harvest mouse lives in a nest like a bird's under tall grass. It builds its nest with stems and leaves.

Meadowlarks also build nests in grass. They cover their nests with grassy roofs. Sometimes these birds build tunnels that lead to their nests. Meadowlarks eat seeds, grasshoppers, and cutworms. Cutworms live in the soil. They eat the roots of plants. Farmers like meadowlarks. Can you guess why?

Den: An animal shelter.

Have you ever heard the howl of a coyote (**ky**-ote)? A coyote looks like a wolf, but it is smaller. Coyotes are hunters that live in the grassland. Coyotes eat mice, rabbits, other small animals, and plants. Coyotes live in **dens**. A *den* is an animal shelter. Coyotes make their dens under the ground or in rocks.

A prairie dog is a small animal that looks a little like a squirrel. Millions of prairie dogs used to live in the grasslands. But now there are not many. They have been hunted and killed by people because they can harm crops. They also eat grass that is needed for cattle and sheep.

Prairie dogs dig burrows in the ground. Their burrows are close to one another. A group of prairie dog burrows is called a town. As many as 1,000 prairie dogs can live in a town. And the town can cover several kilometers. Prairie dogs eat grass, roots, and seeds.

ACTIVITY

What can you learn about a bird's habitat by looking at its nest?

A. Gather these materials: paper and pencil.

B. Look closely at the 2 nests in these pictures.

C. List the materials used to make nest A.

D. List the materials used to make nest B.

1. Which nest was probably made by a chickadee? How can you tell?
2. Which nest was probably made by a meadowlark? How can you tell?

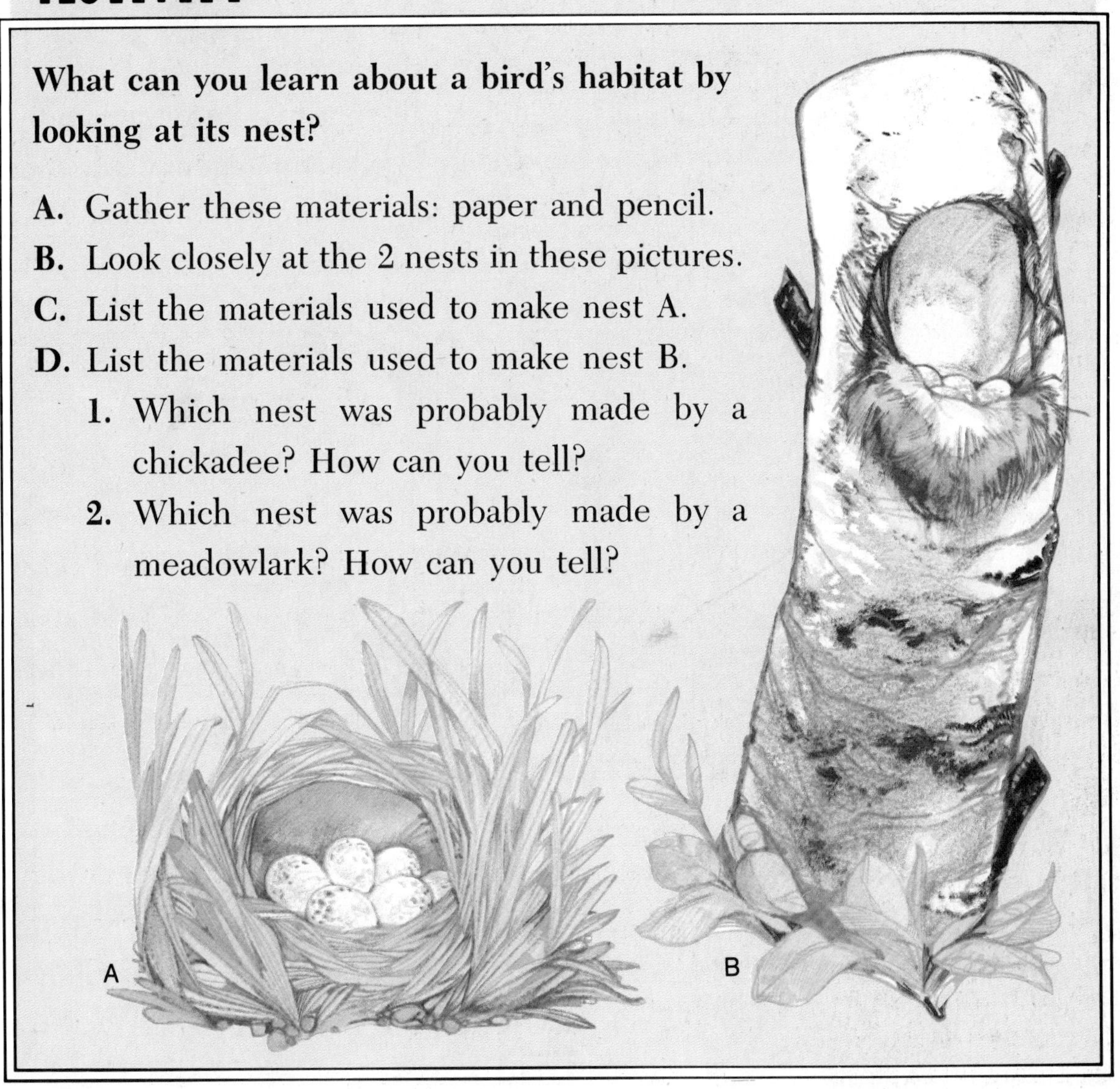

Section Review

Main Ideas: There is not enough rain in the grasslands for trees. But grass and many other plants grow there. In the grasslands, the small animals live in burrows or nests.

Questions: Answer in complete sentences.

1. What is a grassland?
2. What is a prairie dog town?
3. Name three animals that live in a grassland.
4. Where do coyotes make their dens?
5. Where do meadowlarks build their nests?

Jane Goodall

Dr. Jane Goodall studies animals in their habitats. At age eight, Dr. Goodall knew that she would work with wild animals in Africa. There were not many wild animals near her home in England. So she watched the chickens. And she read about animals all the time.

In 1960, Dr. Goodall was invited to Africa. She lived and worked at the Gombe Stream Game Reserve. Her first study was about chimps making tools. Dr. Goodall no longer lives at Gombe. But she travels and gives talks to raise money for work there. Several TV programs have been made about Gombe. Dr. Goodall also teaches at Stanford University.

Chapter Review

Science Words: Match the terms in column A with the definitions in column B.

Column A	Column B
1. Shelter	a. A place with many trees
2. Habitat	b. A place that protects or covers an animal
3. Den	c. A community of different kinds of grasses, wildflowers, and other plants
4. Burrow	d. A safe place for a coyote
5. Forest	e. A place where a plant or an animal lives
6. Community	f. A tunnel under the ground
7. Grassland	g. A group of plants and animals living in one place

Questions: Answer in complete sentences.

1. What is a shelter?
2. What do salamanders eat?
3. What are bison?
4. Where does a harvest mouse live?
5. Why aren't there many trees in grassland areas?
6. Why do deer like to rest in leafy areas of the forest?
7. Why are there different kinds of forests?
8. What do deer eat during the winter?
9. Where do cutworms live?

CHAPTER 17

THE DESERT AND THE TUNDRA

17-1. The Desert

Look at this picture. Have you ever been in a desert? Do you know what kinds of plants and animals live in a desert? Where do these living things get water and food?

When you finish this section, you should be able to:

☐ **A.** Describe a desert.

☐ **B.** Name some plants and animals that live in a desert.

☐ **C.** Tell how these plants and animals find food and water in a desert.

It is very hot in the desert most of the time. The temperature can rise to 38°C (100°F) or more. It is also very dry in desert areas. It rains only a few times each year. Yet large communities of plants and animals live in our deserts.

Most of the plants and animals of the desert are **hardy.** *Hardy* means living under difficult conditions. Some of the plants and animals that live in the desert have special ways to get and to save water. They can live a long time without food and water.

Let's look at a few plants that live in the desert. The mesquite (mes-**keet**) has small leaves and long roots. Under the dry ground, there is water. It is located very deep in the earth. The long roots of the mesquite are able to reach this water. Unlike most plants, the mesquite does not use a lot of water on hot days. Do you know why? Its leaves curl up. When its leaves are curled, the plant is resting. During these times, it uses very little water.

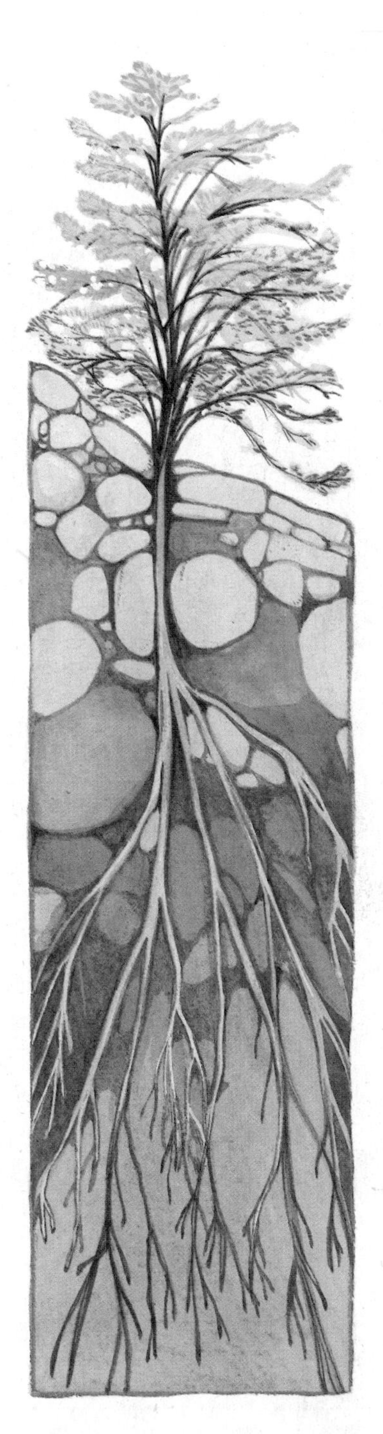

Hardy: Living under difficult conditions.

Spines: The sharp, thorn-like points on a cactus.

Look at the picture above. What are these plants? They are cactuses (**kak**-tus-ez). A cactus is a plant with a thick stem. It has a waxy skin and many **spines**. *Spines* are the sharp, thorn-like points of a cactus.

A cactus has many long roots. The roots spread out around the plant. They grow close to the surface of the ground. When it rains, the roots quickly soak up water from around the plant. This water is kept in the thick stem. By using this water, a cactus can live for a long time.

Now let's learn about some of the animals that live in the desert. How are they able to live where it is so hot and dry?

The picture on this page shows the desert on a hot summer day. On days like this, the desert floor is too hot for most animals. So they seek shelter from the sun. It is cooler under a tree or a rock than in the sun. It is even cooler underground. Where are the desert tortoise (**tor**-tis), the kangaroo rat, and the sidewinder snake in this picture? Can you find the chuckwalla (**chuk**-wal-uh) lizard and the jackrabbit?

Look at the picture again. What animals are moving around? Why is it easier for these animals to move? They have wings, so they do not have to walk on the hot ground. Can you find the Gila (**hee**-luh) woodpecker? It lives in the tall cactus.

This is a picture of the desert at night. It is cooler now. Some animals are sleeping. Others are looking for food and water. Most of them will get enough water from the food they eat.

The kangaroo rat has found some seeds. It carries the seeds in its cheeks. Then it stores the seeds in its burrow. The kangaroo rat never drinks water. It gets its water from seeds. The jackrabbit and the chuckwalla are looking for young leaves. There is water in these leaves. The desert tortoise is looking for some young cactus plants. The tortoise can go a long time without water. Do you know why? Under its shell, it has two bags of skin. In these bags, the tortoise can store enough water to last for many months.

Activity

Is it cooler underground than above ground on a hot day?

A. Gather these materials: 2 thermometers, spoon, and cardboard.

B. Put a thermometer in a hole in the ground. Cover the hole with cardboard.

C. Put the other thermometer on top of the ground. Wait 5 minutes. Then read the temperature on each thermometer.

1. What was the difference between the temperatures of the thermometers?
2. Is it cooler under or above ground?

Section Review

Main Ideas: A desert is very hot and dry. But some hardy plants and animals do live there. They have special ways of getting and keeping water and food.

Questions: Answer in complete sentences.

1. Why doesn't the mesquite use a lot of water on hot days?
2. What is a cactus?
3. What desert animals spend the day underground?
4. Where does the desert tortoise store water?

17-2.

The Tundra

Do you know where the Arctic Ocean is? Find it on a map or globe. There are treeless plains near this ocean and in other arctic areas. It is windy and cold in these places. Very little rain falls. In fact, there is also very little snow. If it does snow, the snow stays on the ground a long time.

Do you know another name for an arctic plain? Can you name some plants and animals that live in this cold habitat?

When you finish this section, you should be able to:

☐ **A.** Give the name for an arctic plain.

☐ **B.** Name some plants that live in this habitat.

☐ **C.** Name some arctic animals and tell how they live.

An arctic plain is called a **tundra** (**tun**-druh). There are no trees on the *tundra*. The topsoil of the tundra is black and mucky. The subsoil is always frozen. Tundras are found in the far northern parts of the world.

In the tundra, winters are long and cold. The ground is frozen. Summers are short and cool. Only the top layer of ground **thaws**. *Thaw* means to become unfrozen. When soil thaws, it becomes damp and heavy.

It is hard for trees to grow in an area where most of the soil is frozen. But there are some grasses, wildflowers, and bushes that can grow in the tundra. Mosses and **lichens** (**ly**-kenz) also grow there. *Lichens* can grow on rocks or rocky ground.

Look at the picture on this page. Most plants in the tundra are small. They grow close to the ground. Why do you think most of the tundra plants grow this way?

Tundra: A grassy arctic plain that is frozen most of the year.

Thaw: To become unfrozen.

Lichen: A plant that can grow on rock or rocky soil.

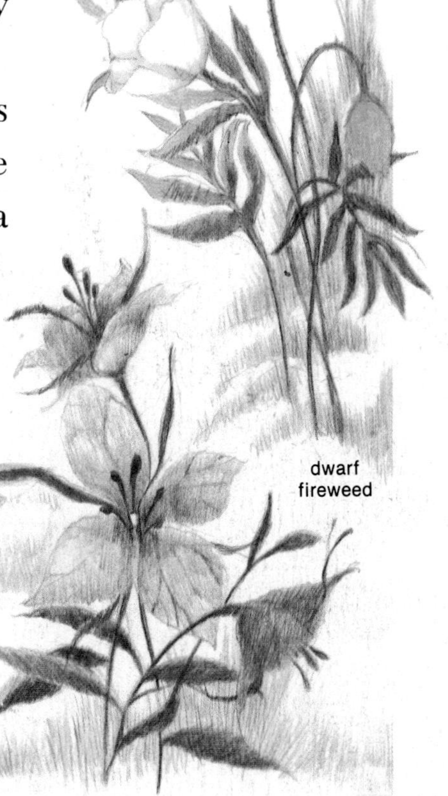

The tundra is too cold for most animals. But there are some hardy animals that live there. Let's learn about some of these animals.

Herd: A large group of animals.

Migrate: To move from one place to another.

The caribou (**kar**-uh-boo) is a type of deer. Look at the caribou in the picture on page 254. Large **herds**, or groups, of caribou roam across the tundra each summer. Because of their wide hoofs, they do not sink into the soggy soil. The caribou come to the tundra to eat grass and lichens. During the winter, the *herds* of caribou **migrate** (**my**-grayt), or move, south. They *migrate* to forests where they can find food.

How do you keep warm when you are outside during the winter? You probably wear a heavy coat. Most arctic animals grow thick coats each winter. The winter coats of some animals are white. Can you guess why? In the summer, the arctic hare has brown fur. But in the winter, its fur is white. The hare lives in the open tundra. It does not make a burrow. During storms, it hides in deep snow. How would its white coat protect it during the winter?

Lemmings are among the most common animals of the tundra. These small animals look like fat mice. Thousands of lemmings live in burrows under the ground. They eat grass, berries, and other arctic plants. When lemmings are out of their burrows, they must watch for foxes. Can you guess why? They are something that foxes like to eat.

Both lemmings and foxes grow thick fur coats in winter. Look at the pictures in the margin. The top picture shows the foot of an arctic fox during the summer. Now look at the bottom picture. It shows the foot during the winter. What happens to the bottom of the foot during the winter?

The ptarmigan (**tar**-mih-gun) must also watch for foxes. Arctic foxes like ptarmigan eggs. A ptarmigan is an arctic bird. Its thick feathers protect it from the cold. During the summer, its feathers are dark. During the winter, they are white. The ptarmigan builds its nest in rocks. Ptarmigans eat mosses, berries, and buds.

Can you find the ptarmigan in these two pictures? How does the bird's color help it in summer? How does its color help it in winter? Suppose you wanted to hide in a forest or grassland. What color would you want your clothes to be? Why?

Section Review

Main Ideas: The tundra is too cold for most plants and animals. Only very hardy plants and animals can live in this arctic plain. Most arctic animals grow thick coats in winter.

Questions: Answer in complete sentences.

1. Where are the tundra areas of the world?
2. Why don't trees grow in the tundra?
3. What is a caribou?
4. What animal likes to eat lemmings and ptarmigan eggs?
5. Why don't caribou sink into the soggy soil of the tundra?

CHAPTER REVIEW

Science Words: Write the sentences below on paper. Fill in the blanks with the correct words from the list.

tundra **migrate** **herds** **thaws** **spines** **hardy**

1. When a plant or animal is tough and can stand bad conditions, we say it is ______.
2. The sharp, thorn-like points on a cactus are its ______.
3. A ______ is a grassy plain that is frozen most of the year.
4. In an arctic plain, only the top layer of soil ______ in the summer.
5. Many large ______ of caribou roam the arctic plain.
6. In winter, these groups of caribou ______ south to find food in forests.

Questions: Answer in complete sentences.

1. What is a desert?
2. Describe a tundra.
3. How are the living things of a tundra like those of a desert?
4. How do animals in the desert keep cool?
5. How do animals in the tundra keep warm?
6. Name two plants that grow in the desert.
7. Where might you find lichens growing?
8. What color are the feathers of a ptarmigan during the winter?

CHAPTER 18

WATER HABITATS

18-1. The Ocean

You are at the sandy edge of the land. The smell of fish and salt is in the air. You see waves and a huge body of water. What is this water? It is the ocean.

When you finish this section, you should be able to:

- ☐ **A.** Explain what an ocean is.
- ☐ **B.** Name some living things that are found in oceans.

An ocean is a huge body of saltwater. Much of the earth is covered by oceans. In fact, there is more water than land on earth.

The ocean is a habitat for many living things. Many types of algae live in the ocean. In Unit 1, you learned that algae are plant-like organisms. Some algae float in the ocean. Others are attached to rocks. Algae do not need soil. But they do need sunlight. Parts of the ocean are very deep. As shown below, the deepest parts of the ocean are very dark. Algae live only in the upper parts of the ocean where there is light.

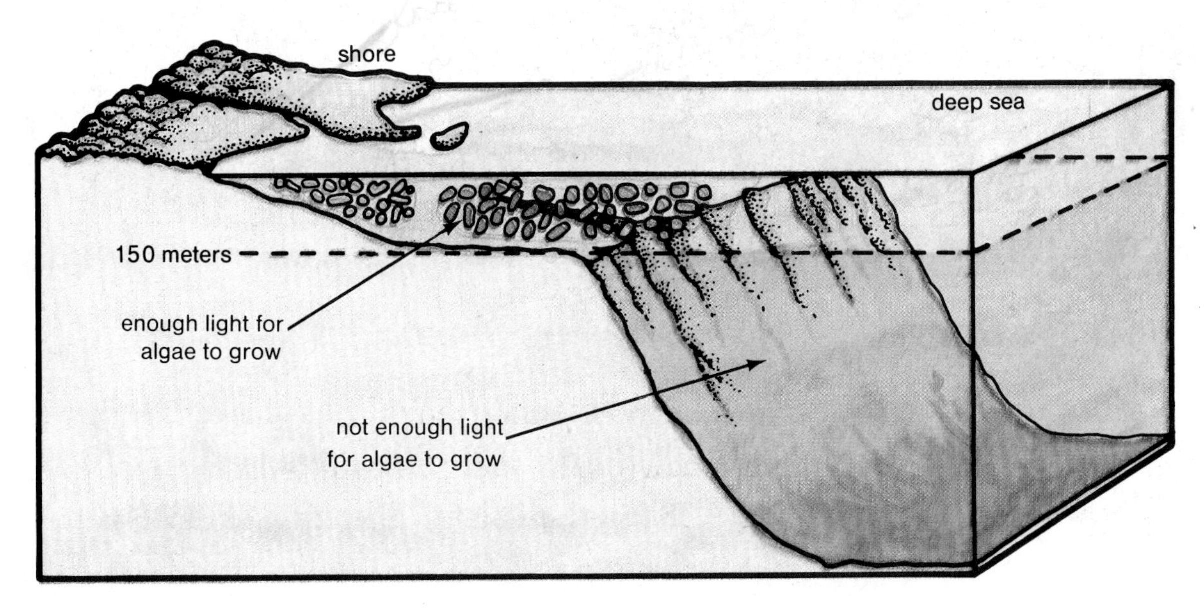

Pretend you are at the ocean. You jump into the water. Something slimy brushes your leg. It looks like a long leaf. What is it? It is probably a type of large algae. Large algae are called seaweed. There are many kinds of large algae. The largest is kelp. Kelp is a type of brown algae. Some giant kelp are over 50 meters (163 feet) long.

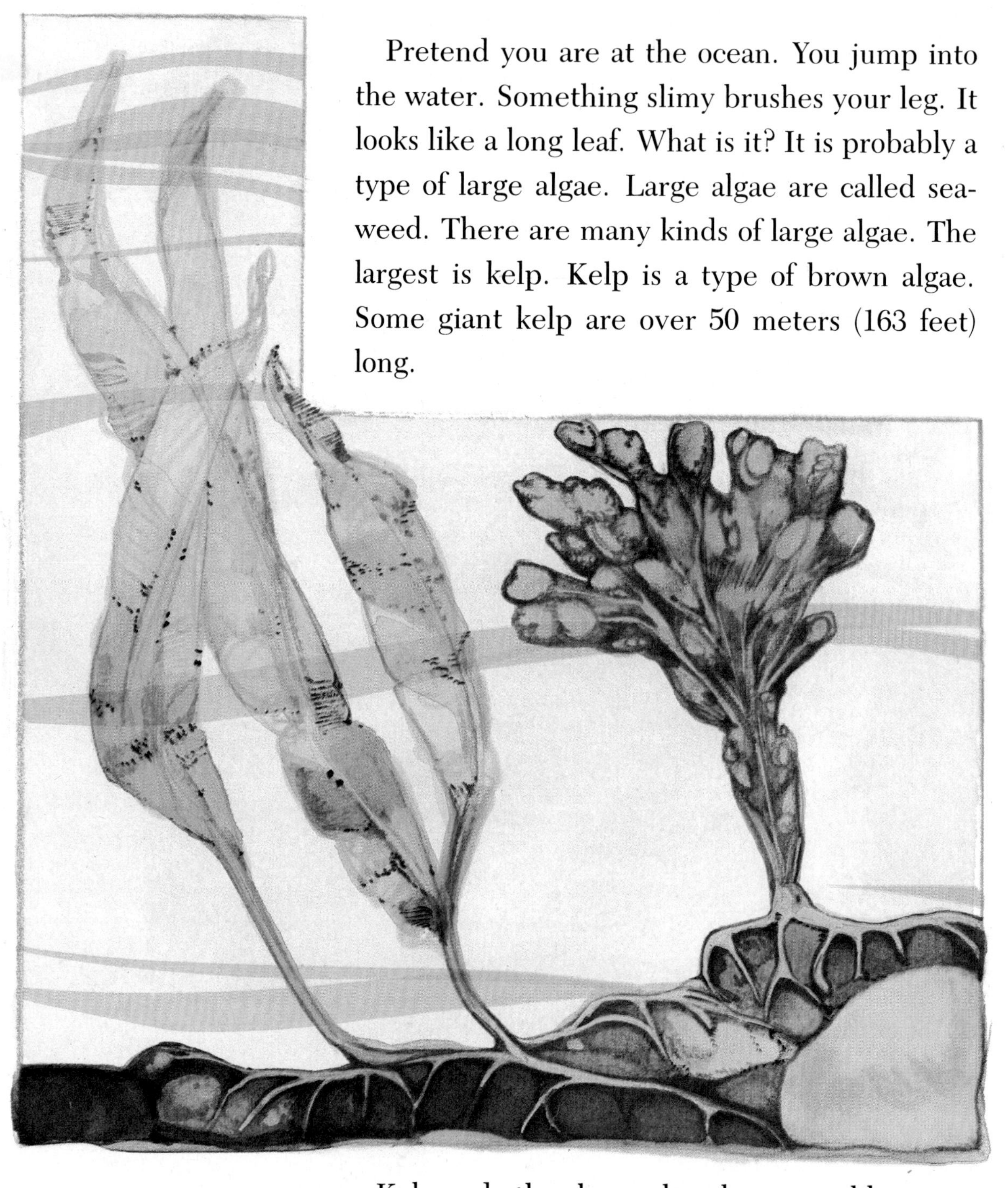

Kelp and other large algae have root-like parts called **holdfasts**. The *holdfasts* hold the plants to rocks. Sometimes large masses of kelp live together. The picture shows some kinds of kelp.

Holdfasts: Root-like parts of algae.

Many algae are too small to be seen without a microscope. Sometimes billions of algae form masses of green near the surface of the ocean.

Tiny algae and animals that live in the ocean are called **plankton**. The picture below shows some *plankton*. The shrimp-like animals are krill. Some of the algae are diatoms (**dy**-uh-tahmz). Many fish eat plankton. So does the largest animal on earth. Can you name this animal?

Plankton: Algae and small animals that float in the ocean.

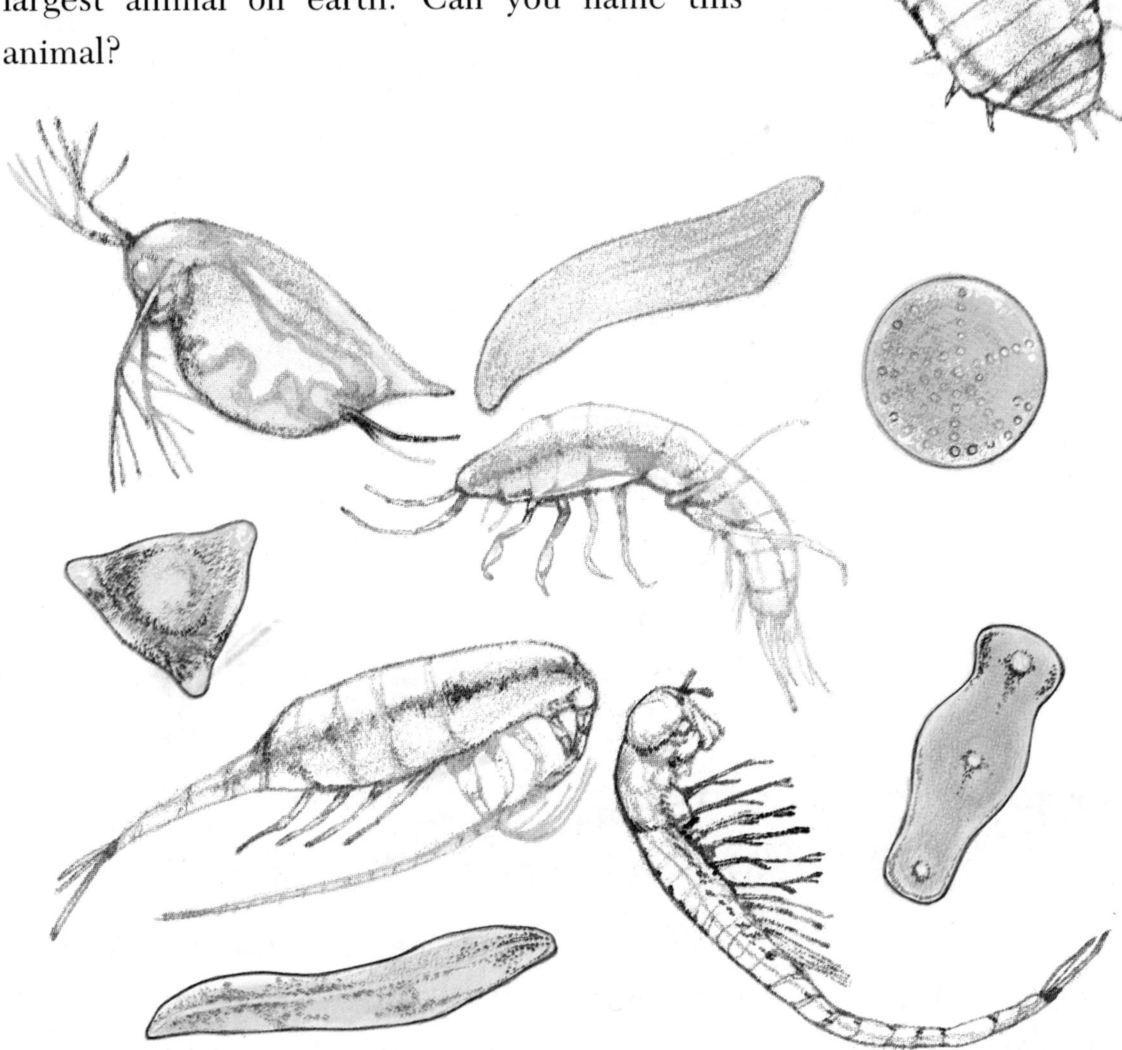

The blue whale is the largest animal on earth. It is many times larger than an elephant. The blue whale migrates across the oceans of the world each year. It looks for plankton. In summer, the blue whale is in the Arctic Ocean. In winter, it feeds in the warm ocean waters farther south.

Baleen: A plate that hangs from a whale's jaw; it strains plankton from the water.

A blue whale does not have teeth. Instead, **baleen** (buh-**leen**) plates hang from its upper jaw. The *baleen* acts like a strainer. When the whale is hungry, it takes in large amounts of ocean water and plankton. The baleen plates strain out the plankton. During its life, a blue whale eats tons and tons of plankton.

School: A group of fish that swim together.

Many fish also eat plankton. Most fish eat plankton only when they are young. But herrings (**hair**-ings) eat plankton even as adults. Herrings are a very common fish. Large **schools**, or groups, of herrings swim in all the oceans.

Most ocean animals live in the upper parts of the ocean. Do you know why? There is lots of food in these places.

One of the strange animals that live in the cold, dark part of the deep ocean is shown on page 266. It is called an anglerfish. This fish can move a long piece of its top fin so that it hangs in

ACTIVITY

Can plants live without sunlight?

A. Gather these materials: water that has been standing overnight, 2 glass jars, and 2 pieces of water plant.

B. Fill the glass jars with the water. Place 1 piece of the water plant in each jar.

C. Set 1 jar in a sunny window.

D. Set the other jar in a dark closet.

E. After 15 minutes, look at the plant in the window. Do you see bubbles coming from the cut end? Look at the plant in the closet. Do you see bubbles coming from it? When a plant has bubbles coming from it, it usually means that the plant is healthy.

1. What effect did the sunlight have on the plant?
2. Do plants need sunlight?

front of its mouth. The end of this piece glows. Other fish notice it. When another fish comes close to the glowing fin, the anglerfish quickly eats it.

Section Review

Main Ideas: There are many living things in the ocean. Many kinds of algae and small animals live in the upper part of the ocean. Fish and larger animals feed on the tiny plants and animals.

Questions: Answer in complete sentences.

1. What is an ocean?
2. Where does kelp grow?
3. Where could you find large masses of algae?
4. Where do anglerfish live?
5. Name a large animal that eats plankton.
6. What is the largest kind of algae?
7. What is the largest animal on earth?

18-2.

The Shore

The edge of the ocean where the water meets the land is called the shore. Some shores are rocky. Others are sandy. Many plants and animals live at the shore. Those that live at the water's edge must be very hardy. Can you guess why?

When you finish this section, you should be able to:

- ☐ **A.** Describe two kinds of shores.
- ☐ **B.** Name some plants and animals that live at the shore.
- ☐ **C.** Explain why only hardy plants and animals can live at the water's edge.

High tide: When water moves far, or high, onto the shore of an ocean.

Low tide: When water stays low on the shore.

Compare the picture below to the one on page 267. How are these places alike? How are they different?

Both pictures show the same rocky shore. The picture on page 267 was taken at **high tide**. The picture on this page was taken at **low tide**. At *high tide*, the water moves up high onto the shore. At *low tide*, the water stays low.

Many living things are found at the shore. They must be hardy because of the waves that hit the shore. They must also be able to live out of water as well as underwater. At high tide, they are underwater. At low tide, they are out of water.

Did you notice the brown, slimy algae growing on the rocks in the picture on page 268? This is rockweed. Rockweed is a hardy algae. The pounding waves do not harm it. At low tide, rockweed begins to dry out in the sun. At high tide, it becomes wet again.

Some animals with shells cling to rocks and other solid objects at the shore. One of these animals is the barnacle (**bar**-nuh-kul). The barnacle has a soft body. It lives on its back inside its shell. At high tide, it opens its "door" and sweeps plankton into its shell. At low tide, it closes its "door."

The strange creature with a petal-like top is a sea anemone (uh-**nem**-oh-nee). The petals are called **tentacles** (**ten**-tuh-kulz). When it is covered by water, the sea anemone moves food into its mouth with these *tentacles*. The anemone can live in the waves. It holds onto rocks with a part of its body called a "foot."

Tentacles: Petal-like parts of the sea anemone that are used to get food.

The living things at a sandy shore have special problems. Can you guess what these problems are? They have nothing to hide behind. There is nothing to hang on to. For this reason, not many plants or large algae live at a sandy shore. But some small, hardy animals do live there. If you dig into the sand, you will find them. Most of the animals dig burrows into the sand.

One of these animals is shown in the picture at the upper left. This is a mole crab. It takes the crab only a few seconds to burrow out of sight. Many other shelled animals also burrow into the sand. Some kinds of clams live deep in the sand. At low tide, the clams stay hidden. At high tide, they move up through the sand. Why do you think they do this? It lets them take in water and plankton.

Ocean birds do not build nests at the edge of the water. But many birds look for food near the water. Look at the picture on page 271. This is a sandpiper. You can find this bird at many shores. It runs back and forth in front of the waves. It is looking for small animals that are washed out of their burrows by waves. Some waves bring in small animals from the nearby ocean floor.

Section Review

Main Ideas: At high tide, ocean water moves up high onto the shore. At low tide, it stays low on the shore. On a sandy shore, living things must be able to live both underwater and out of water.

Questions: Answer in complete sentences.

1. Describe two kinds of shores.
2. Where is rockweed found?
3. Why do animals at a sandy shore dig tunnels?
4. How does a barnacle take in plankton?
5. Why must living things at a shore be hardy?
6. Draw a picture of a sea anemone. Label your picture.

18-3.

The Pond

Do you know what the plants in the picture below are? Where do these plants grow? The plants are water lilies (**lihl**-eez). They often grow in ponds.

A pond is a small body of water. Most ponds are not very deep. A large community of plants and animals can live in a pond. A pond is their habitat. The pond has everything they need to live.

When you finish this section, you should be able to:

☐ **A.** Describe a pond.

☐ **B.** Name some water plants and tell where they grow.

☐ **C.** Name some pond animals and describe how they live.

The picture on this page shows some of the many kinds of water plants and where they grow. Some plants grow near the shore. They can grow even in water that is not very deep.

Look at the picture. Can you find the cattails and the arrowheads? These plants grow near water. Cattails have long leaves and brown, fuzzy flower spikes. The leaves of arrowheads are pointed like the heads of arrows. They have white flowers. Can you find the iris in the picture? It has big blue flowers.

Other plants seem to float on the water. The water lily grows this way. Actually, only its leaves and flowers float. The long stem of the plant reaches to the mud at the bottom of the pond. Its roots hold the plant in place. Tiny plants, like the duckweed, live on top of the water.

Many kinds of algae also live in a pond. Most are too small to be seen without a microscope. They float in the water. As shown on pages 43 and 274, some algae can form a green scum.

Where do frogs live? Many live in ponds. Have you ever heard this sound: *jug-er-rum*? It is the sound of a bullfrog. A bullfrog is a very large frog. Some bullfrogs are over 20 centimeters (8 inches) long. A bullfrog eats insects, fish, other frogs, and birds. Young frogs and tadpoles eat algae, other water plants, and small animals.

In spring and summer, bullfrogs live in water at the edge of a pond. Before winter comes, the bullfrog digs a burrow in the mud at the bottom of the pond. It sleeps in the burrow all winter. The burrow keeps the bullfrog warm.

Can you name some other animals that live in ponds? Some of them are shown in the picture on page 275. The insect with four wings is a dragonfly. The beautiful gray or brown insects are mayflies. The brownish black one is a diving beetle. This one has light markings. When it is out of the water, the diving beetle breathes through openings in its body. When it dives to the bottom of the pond, it carries a bubble of air between its legs. This lets it stay underwater a long time.

Do you know what these insects eat? They eat other bugs, tiny fish, and other small water animals. In turn, they are eaten by bullfrogs or by large fish.

Two fish that live in ponds are the yellow perch and the bullhead. Can you find these fish in the picture? The large bird is a blue heron (**hair**-on). As you can see, it comes to the pond to grab fish with its long beak. Many turtles live in ponds. Like frogs, they sleep in mud during the winter.

ACTIVITY

Can you make a water habitat?

A. Gather these materials: jar or aquarium, clean gravel, rocks, water, water plants, 2 snails, fish, and fish food.

B. Put 3 to 5 centimeters of gravel and some rocks on the bottom of the jar.

C. Slowly fill the jar with water.

D. Plant the water plants in the gravel. Then, lower the snails and fish into the water.

E. Feed the fish each day. Keep the habitat clean. Record any changes you see in the habitat.

Section Review

Main Ideas: A pond is a habitat for many living things. Plants live in different parts of the pond. Animals find many things to eat in ponds.

Questions: Answer in complete sentences.

1. What is a pond?
2. Name two plants that live near water.
3. Name two things that bullfrogs eat.
4. What do dragonflies and diving beetles eat?
5. Describe a water lily.
6. Where do bullfrogs and turtles live during the winter?

Chapter Review

Science Words: The clues in column B will help you unscramble the words in column A.

Column A	Column B
1. HGIH DIET	When water moves high onto the shore
2. TNOPKLAN	Algae and small animals that float in the ocean
3. DLOSTFHAS	Root-like parts of algae
4. NAEBEL	Plate that hangs from a whale's jaw
5. SLEECTNAT	Petal-like parts on an animal that are used to get food
6. LOHSCO	A group of fish that swim together

Questions: Answer in complete sentences.

1. Name three kinds of water habitats.
2. Where does the blue whale look for food during the summer?
3. What is the difference between high tide and low tide?
4. What is another name for a large group of fish?
5. Draw a picture of a cattail.
6. What are holdfasts?
7. Name two animals that live in burrows in the sand.
8. How does a whale use its baleen plates?

INVESTIGATING

What lives in a habitat near you?

A. Gather these materials: small notebook, pencil, spoon, newspaper, plastic bags, and string.

B. Visit a grassy area, a forest, or an open place.

C. Using the string, mark off a small area about 1 meter square.

D. Look at the area very closely.

- **1.** What kind of plants do you find?
- **2.** What kind of animals do you find?

E. Write your findings in the notebook. Draw pictures of the plants and animals.

F. Using the spoon, dig up some soil from the area.

G. Put the soil on the newspaper. Look at it closely.

- **3.** Did you find any animals in the soil?
- **4.** Describe the soil.

H. Write your findings in your notebook.

I. Perhaps you would like to study some worms or insects more closely. Put them in plastic bags and take them back to class. Be sure to include some soil and plant material with the animals.

CAREERS

Zoologist ▶

You have learned how animals live together. Zoologists (zoo-**ahl**-uh-jists) study the different parts of each animal's body. They also study how animals live together. Some zoologists study how animals adapt to sudden changes, such as very cold weather. People who are zoologists have studied biology.

◀ Fishery Manager

You have learned about water habitats and some of the fish that live in ponds and oceans. A fishery manager works to provide proper habitats for fish. Fisheries raise fish to be put into lakes and rivers. Some college training is needed if you wish to work as a fishery manager.

GLOSSARY/INDEX

In this Glossary/Index, the definitions are printed in *italics*.

Air, 137–140
Algae (**al**-jee): *plant-like organisms with chloroplasts*, 44–47, 55, 262–263, 266; in ponds, 273, 274; on shores, 269, 270
Allosaurus (**al**-uh-**sawr**-us): *a meat-eating dinosaur*, 94
Ameba (uh-**mee**-buh): *a type of protozoan that has no real shape*, 40–41
Anglerfish, 265–266
Animals, arctic, 256–258; compared to plants, 13; desert, 249, 251–252; development of, 28, 30; food important to, 9; fur coats of, 256, 257, 258; grassland, 241–246; ocean, 264–266; pond, 274–275; reproduction of, 10; shore, 269, 271
Arctic (**ahrk**-tic): *a very cold habitat*, 254
Arrowheads, 273
Astronaut (**as**-troh-naut): *a person who travels in space*, 156, 157
Astronomer: *a scientist who studies the solar system and other objects in the sky*, 189
Atmosphere (**at**-mus-feer): *the gases that surround a planet*, 184
Axis (**ak**-sis): *an imaginary line running through the earth from the North Pole to the South Pole*, 149

Bacteria (back-**teer**-ee-uh): *very small one-celled living things*, 39–42, 55
Baleen (buh-**leen**): *a plate that hangs from a whale's jaw, straining plankton from water*, 264
Barnacle (**bar**-nuh-kul): *a shore animal that has a soft body*, 269
Bedrock: *the bottom layer of soil*, 75
Beetles, 236, 274–275
Big Dipper: *a constellation that looks like a big spoon*, 169, 170, 171
Bionic: *something that uses electricity to help the body work*, 228
Birds, 270–271
Bison (**bye**-son): *American buffalo*, 241
Boil: *to change quickly from a liquid into a gas*, 109
Boiling point: *temperature at which matter boils*, 128
Botanist: *a person who studies plant life*, 55
Brontosaurus (**brahn**-tuh-**sawr**-us): *a plant-eating dinosaur*, 93–94
Buffalo (**buh**-fuh-loh): *American bison*, 241
Bullhead, 275
Burrow (**buh**-roh): *a tunnel under the ground*, 238, 245, 270, 274

Cactus (**kak**-tus): *a plant with a thick stem*, 250
Carbon dioxide (**kahr**-bun dy-**ahk**-side): *a gas in the air*, 47, 72, 179
Caribou (**kar**-uh-boo): *a type of deer that roams across the tundra in the summer*, 256
Cattails, 273
Cell(s): *a small unit of life*, 14–18, 26–29, 37–42, 43–46, 49, 51; energy needs of, 23, 46–47; of plants compared to animals, 18; reproduction of, 22–25; types of, 32–35

PHOTO CREDITS

HRW Photos by Bruce Buck appear on pages 208, 209.
HRW Photos by Russell Dian appear on pages 49, 65 bottom right, 76 bottom, 80 top, 104, 105, 106, 107, 108, 109, 110, 112 top, 113 top, 114 top, 117, 123 all, 127, 130, 134, 135 top, 139 top, 149, 194 bottom, 196 both, 199, 204 bottom, 210, 212 bottom, 216, 217, 231 top, 260, 267, 268, 270 top.

HRW Photo by Roland Cormier appears on page 151.

HRW Photos by Louis Fernandez appear on pages 11 left, 23, 26, 48, 67 left, 71 bottom, 79, 121, 138, 193, 195 both, 211, 212 top, 221, 265.

HRW Photo by Curt Gunther appears on page 8.

HRW Photo by Richard Haynes appears on page 205.

HRW Photos by Richard Hutchings appear on pages 12 top, 102, 114 bottom.

HRW Photos by Ken Karn appear on pages 20, 21, 31, 60 bottom, 62, 63, 65 top, 66, 72, 73 bottom, 77, 98, 119, 154, 164, 185, 218.

HRW Photo by David Spindel appears on pages 190–191.

HRW Photos by Yoav Levy/Phototake appear on pages 14, 27, 113 bottom, 116, 124 top, 129, 135 bottom, 136, 139 bottom, 142, 192, 201, 206, 209 middle, 219, 226, 230.

p. iv top—Brack Kent/Earth Scenes, bottom—J. Freilich; p. v top—John Zoiner/Peter Arnold, Inc., bottom—NASA; p. vi—L. Brown/The Image Bank; p. vii—Robert Ashworth/Photo Researchers, Inc., p. viii—H. S. Benton, M.D./Photo Researchers, Inc.; p. 1—Animals Animals/Oxford Scientific Films.

Unit 1: pp. 6-7—Kojo Tanaka/Animals Animals; p. 10 top—Leonard Lee Rue III/Bruce Coleman, Inc., bottom—Jeff Foott/Bruce Coleman, Inc.; p. 11 right—Leszczynski/Earth Scenes; p. 12 both—Wil Blanche/D.P.I.; p. 15—Runk/Schoenberger/Grant Heilman; p. 18—Morton Beebe/The Image Bank; p. 24 left— Tom McHugh/Photo Researchers, Inc., right—R. Bornemann/Photo Researchers, Inc.; p. 25—University of Santa Barbara; p. 33—Manuel Rodriguez; p. 34 left—Raymond A. Mendez/Animals Animals, right—Zig Leszczynski/Animals Animals; p. 37—Breck Kent/Earth Scenes; p. 38—R. U. Utlman/Taurus Photos; p. 42—Kenneth Fink/Bruce Coleman, Inc.; p. 43—Harry Engles/Earth Scenes; p. 44 top—Runk/Schoenberger/Grant Heilman, bottom—Bob Evans/Peter Arnold, Inc.; p. 45 top left—Manfred Kage/Peter Arnold, Inc., top right—W. H. Hodge/Peter Arnold, Inc.; bottom left—E. R. Degginger, bottom right—Manfred Kage/Peter Arnold, Inc.; p. 47—Grant Heilman; p. 52 both—W. H. Hodge/Peter Arnold, Inc., p. 55 top—Wil Blanche/D.P.I.

Unit 2: pp. 56-57—Jo Snyder; p. 58—Alex Stewart/The Image Bank; p. 59 left—Alex Stewart/The Image Bank, right—Nick Nicholson/The Image Bank; p. 60 top—Gary Brash; p. 61 top and middle left—Russ Kinne/Photo Researchers, Inc., middle right—George Whitely/Photo Researchers, Inc., bottom left—John Neel/Photo Researchers, Inc., bottom right—Tom McHugh/Photo Researchers, Inc.; p. 64 top—American Museum of Natural History, bottom—Harold Sund/The Image Bank; p. 65 bottom left—Tiffany & Co.; p. 67 right—Aram Gesar/The Image Bank; p. 69—A. Blank/Bruce Coleman, Inc.; p. 70—Stouffer Production Ltd./The Image Bank; p. 71 top—H. M. Duke Schoonouer/The Image Bank; p. 73 top—George Remain; p. 74—Nicki Marechal/The Image Bank; p. 75—USDA Soil Conservation Service; p. 76 top—Grant Heilman; p. 78—Farrell Grehan/Photo Researchers, Inc.; p. 80 bottom left—Susan McCartney/Photo Researchers, Inc., bottom right—Thomas Brave/The Stock Market; p. 83—J. Freilich; p. 84—Townsend P. Dickenson/Photo Researchers, Inc.; p. 85 top right—Thomas Martin/Photo Researchers, Inc., middle—George Gerster/Photo Researchers, Inc., bottom right—Louise K. Broman/Photo Researchers, Inc.; p. 87—Mark Hanaur; p. 88—Dr. Schwimmer/Bruce Coleman, Inc.; p. 93 left—Dinosaur State Park, Rocky Hill, Conn., right—American Museum of Natural History; p. 99 right—Mark Hanaur, left—Charles B. Ford.

Unit 3: pp. 100-101—Hank Morgan/The Image Bank; p. 111—Fournier/Rapho/Photo Researchers, Inc.; p. 115 top—Bullaty/Lomeo, bottom—Courtesy George Mason University; p. 118—Dan McCoy/Rainbow; p. 122—Grant Heilman; p. 124 bottom—Grant Heilman; p. 125—Dan McCoy/Rainbow; p. 128—Dick Don/D.P.I.; p. 133—John Zoiner/Peter Arnold, Inc.; p. 137—Wil Blanche/D.P.I.; p. 140—Pam Hasegawa/Taurus Photos; p. 143 bottom—Dan McCoy/Rainbow, top—Bohdan Hrynewych/Stock/Boston.

Unit 4: pp. 144-145—NASA; p. 146—NASA; p. 156—NASA; p. 157 top and bottom—NASA; p. 158 top—NASA, bottom—Peter Bloomer/Horizons West; p. 159—NASA; p. 160—L. Wang/Contact; p. 162—Naval Research Lab/NASA; p. 166—Dennis Milton; p. 167—U.P.I.; pp. 173, 177, 178, 179, 180, 182, 183, 184—NASA; p. 189 top—Cerro Tololo/Inter-American Observatory, bottom—Azann/Black Star.

Unit 5: p. 202—L. Brown/The Image Bank; p. 220—Wang, Inc.; p. 222—J. Pickerell/Black Star; p. 223 right—J. Nacktwey/Black Star, left—T. McHugh/Photo Researchers, Inc.; p. 225—M. Manheim/Photo Researchers, Inc.; p. 227 left—M. Abbey/Photo Researchers, Inc., right—Dan McCoy/Black Star; p. 228 left—Courtesy of Intermedics Inc., right—Dan McCoy/Rainbow; p. 231 bottom—Sepp Seitz/Woodfin Camp & Assoc.

Unit 6: pp. 232-233—Ralph Reinhold/Animals Animals; p. 234—E. R. Degginger/Bruce Coleman, Inc.; p. 236 top left—Joe Devenney/The Image Bank, bottom—Stephen Dalton/ Photo Researchers, Inc., top right—John Lewis Stage/The Image Bank; p. 237 top right—John M. Burney/Bruce Coleman, Inc., bottom—Leonard Lee Rue III/The Image Bank; p. 238—Waldene Weisser/Bruce Coleman, Inc.; p. 240—Zig Leszczynski/Animals Animals; p. 241—Alford Cooper/Photo Researchers, Inc.; p. 243 top right—Bob Gossington/Bruce Coleman, Inc., bottom left—Tom McHugh/Photo Researchers, Inc., bottom right—Wardene Weisser/Bruce Coleman, Inc.; p. 244 top—Simon/Photo Researchers, Inc., bottom—Richard Weiss/Peter Arnold, Inc.; p. 246—Bob Strong/*The New York Times;* p. 248—Robert J. Ashworth/Photo Researchers, Inc.; p. 254—Stephen J. Krasemann/Peter Arnold, Inc.; p. 256—Tom McHugh/Photo Researchers, Inc.; p. 257 left—Michael Lugue/Photo Researchers, Inc., right—Charles Ott/Photo Researchers, Inc.; p. 258 top left and right—Jen and Des Bartlett/Bruce Coleman, Inc.; p. 264—Russ Kinne/Photo Researchers, Inc.; p. 266—Tom McHugh/Photo Researchers, Inc.; p. 269 top right—W. H. Hodge/Peter Arnold, Inc., middle right—Robert Dunne/Photo Researchers, Inc., middle left—E. R. Degginger/Bruce Coleman, Inc.; p. 270 middle—Jen and Des Bartlett/Bruce Coleman, Inc., bottom—New York Zoological Society; p. 271—Leonard Lee Rue III/The Image Bank; p. 272—Grant Heilman; p. 274—E. R. Degginger/Bruce Coleman, Inc.; p. 279 top—New York Zoological Society, bottom—E. R. Degginger/Bruce Coleman, Inc.

Art Credits

Anthony Accardo, page 103

Gary Allen, pages 75, 79, 83, 197, 199, 207, 261

Helen Cogancherry, pages 3, 4, 5, 16, 126, 131, 150, 153, 159, 171, 203, 213, 214, 239, 253, 276, 278

William Hunter-Hicklin, page 188

Marion Krupp, pages 9, 28, 29, 32

Network Graphics Studio, pages 15, 17, 22, 33, 38, 40, 41, 46

Tony Rao, pages 88, 89, 91, 92, 93, 94, 95, 96

Jim Racioppi, pages 235, 242, 255

Joel Snyder, pages 2, 245, 249, 250, 251, 252, 257, 262, 263, 275

Pat Stewart, pages 29, 46, 50, 51, 54, 81, 86, 90, 194, 215, 224

John Kier, pages 147, 148, 149, 152, 155, 157, 163, 167, 168, 169, 170, 174, 175, 177, 185

Vantage Art, Inc., pages 3, 131